Ferguson's (Beauregard's) SC Artillery Company

Farm Lake Press

Ferguson's (Beauregard's) SC Artillery Company

Christopher D. Rucker

ISBN 978-1-7361727-0-4

Typeface: Baskerville Old Face

Cover: Thomas B. Ferguson, *The Citadel Archives and Museum*

Farm Lake Press
329 Farm Lake Road
Boiling Springs, SC 29316
864-680-0897

farmlakepress@outlook.com

Contents

Part I. Eighteen Months to Prepare

The origins of Ferguson's Artillery Company (FAC) must be considered in light of preceding events. South Carolina, waggishly described by James Petigru as "too small for a republic, but too large for an insane asylum," had declared itself an independent nation in 1860, and played an outsized role in the beginning of the Civil War. It was in Charleston that the first shots were fired, whether one argues that they were directed at the steamship *Star of the West* in January of 1861, or later during the April bombardment of Fort Sumter. The men of the Palmetto State responded to the call for volunteers, and by the end of 1861 had seen a year of service on multiple fronts. That first year had seen the southern forces hold their own with significant victories over their northern enemy, and belied the early expectations of a short conflict.

Creating a company of artillery in 1862 forced attention to some realities. The first flush of patriotic spirit had driven the enthusiastic volunteers into their units early on, and recruitment in 1862 would have to be of those men who had initially chosen to sit out the war. Competition between recruiters for these reluctant men meant that officers had to be energetic, persuasive, and sometimes innovative to fill their companies. At the same time, the South's meager industrial base and the Federal blockade of its ports brought logistical challenges to the creation of new military units, which necessarily had to compete with existing field forces for the limited supply of weapons and material.

The creation of Ferguson's as an effective fighting unit is the story of its officers, and their efforts to recruit, outfit, supply, and train men who mostly had never been far from their rural homes, were largely illiterate, unused to military discipline, and susceptible to disease and the temptations of urban living. The successful establishment of FAC over an eighteen month period is a testament to the abilities of its officers, and to the commitment of its soldiers.

Company Flag, author's photo. *SC Confederate Relic Room and Military Museum*

There are two extant flags with connections to Ferguson's Artillery Company. The first is a Charleston-issue, Army of Northern Virginia style artillery-sized battle flag, captured November 26th, 1863 by the 69th Ohio Vols. at Graysville, GA, after the battle of Missionary Ridge. It is in the collection of The American Civil War Museum in Richmond, VA, catalog number WD 090 (the banner) and no. WD 090a (the flagstaff). An image of the tattered banner is available in the museum's online searchable collection. This three foot square flag would have been issued to the company in the spring of 1863, just before it left Charleston for field operations.

The second banner is the one pictured above, in the collection of the South Carolina Confederate Relic Room and Military Museum in Columbia. The purchased flag came with no provenance. A textile consultant's opinion, noted but not available at the museum, was that the flag is a reproduction. To this author's personal examination of the flag, the construction does not suggest that it was either a carefully crafted wartime presentation flag or a post-war creation. It is a rather amateurish job, and is more suggestive of a workmanlike creation such as would have been made in the field by a soldier with limited access to materials. That its text is "Ferguson's Battery" also supports a wartime origin. The wounded Captain Ferguson left the company during its first battle, and after Lt. Beauregard assumed command, the unit was called "Beauregard's" as frequently as "Ferguson's" for the remainder of the war. A reunion flag commemorating the battle history of the company would more likely carry the name of the eldest son of the famous Gen. P.G.T. Beauregard, who led the men until the end.

1. Captain Thomas B. Ferguson

Fig. 1. Cadet Ferguson, Class of 1861. *The Citadel Archives and Museum*, Charleston, SC.

Born in 1841 in St. John's Berkeley Parish, SC, the third child in a prominent, wealthy and influential plantation family, Thomas Barker Ferguson had three brothers who also served in the Confederate forces: Brig. General Samuel Wragg; Maj. J. Dugué; and Sgt. Joseph Sanford of the Marion Artillery, who died in service in June 1863. Thomas attended Charleston's Citadel Military Academy (Fig. 1). Anticipating the onset of hostilities, on January 1st, 1861 a detachment of Citadel cadets including Ferguson was deployed to Cummings Point on Morris Island to construct and man a battery of four 24 pounder guns (later named the Star of the West battery) to defend Charleston Harbor's ship channel. They labored there for a week, and earned their place in history when the *Star of the West* steamed into the channel on January 9th to resupply the beleaguered Fort Sumter.[1] The cadets fired the first shots of what would become the Civil War, and Ferguson pulled the lanyard for the battery's third shot, striking the ship and forcing its retreat.[2]

Ferguson was eligible for commission as an officer due to membership in the Battalion of State Cadets, and after graduation from the Citadel, he served as Maxcy Gregg's adjutant in his 1st SC Volunteers, a six month unit which mustered out in late July.[3] Ferguson's appointment as a captain in the SC Artillery came January 1st, 1862, and he employed considerable talents in the recruitment, organization and training of his battery of light artillery. It was raised as part of the state of South Carolina's regular army, and was transferred into Confederate States service on April 4th, 1862. Recruiting took much of Ferguson's time in the first few months of the organization. Equipping the company posed some challenges, as shortages often meant that his men had to do without some essentials. As a Citadel graduate, he was sensitive to military orders, and did his best to fulfill the requirements of his battery. For example, he acknowledged the Confederacy's standards for uniforms, and obtained red cloth to fashion jacket collars and cuffs in the color prescribed for artillerymen. He requisitioned artillery caps and jackets, and blue trousers, to conform to published standards. Though such articles were not always available, Ferguson knew what his men were supposed to wear, and at least early on did his best to make them fit the bill.

Ferguson was Maj. Gen. William H.T. Walker's artillery chief during the campaign around Jackson, MS. At Jackson on July 14th,

1863, while executing an order from Walker, Ferguson was shot with a Minié ball through the upper lobe of his right lung, surviving but rendered "hors de combat." Such a wound might have been expected to be mortal, and when word of the event was received back in Charleston, family and acquaintances assumed the worst and prematurely mourned his death.[4]

Not until the following spring was the wound healed, and Ferguson needed another two months of garrison duty before he would even be considered for further field command. He therefore requested assignment to the Ordnance Department in order to improve himself in that branch of the service. Ferguson served in the Office of Inspector of Laboratory at the Charleston Arsenal until August of 1864. He was then promoted to major, recommended by Gen. States Rights Gist as "an officer of education, high capacity, unblemished reputation and in every aspect worthy of promotion." Gen. Walker said that Ferguson was "brave, intelligent and enterprising. He is a soldier of high toned, gentlemanly principle and well acquainted with his profession."

Maj. Ferguson served as an Inspector of Artillery in the Dept. of SC, GA and FL for the remainder of the war. He ran for the SC state house as representative from the parishes of St. Phillips and St. Michaels in the November, 1864 election but did not win a seat.[5]

Twenty-four years old at war's end, Ferguson remained in Charleston to attend to plantation business, but eventually saw his future elsewhere. He moved to Baltimore in 1867, and married Jane Byrd Swann of Maryland, the governor's daughter. By education Ferguson was an engineer, and he held an interest in aquaculture. He published a monograph titled *Pisciculture* in 1880, proposing methods to increase the artificial hatching of fish eggs. He was appointed to Maryland's State Fisheries Commission, and served as assistant commissioner of Fish and Fisheries of the United States until 1887.[6] His social background and family ties served him well, earning him an appointment as Envoy Extraordinary and Minister Plenipotentiary to Sweden, serving as ambassador to Sweden and Norway from 1894 until 1898, when he returned from Stockholm to the US.[7] (Fig. 2).

Ferguson moved to Washington, DC, residing in the area of Lafayette Park, very close to the US patent office.[8] He was somewhat of an inventor, having patented an improved coffee maker and proposed improvements in fish farming.[9] Ferguson continued to travel, both back to his childhood home in Charleston, and to see his daughter in Boston. (Fig. 3). He died there while on a visit, August 11th, 1922, at the age of eighty-one. He is buried beside his wife in Baltimore's Green Mount Cemetery.[10]

Fig. 2 Ambassador Ferguson. *LSU Library.*

Fig. 3 Ferguson in his later years. *LSU Library.*

2. The Officers

René Toutant Beauregard
Times Picayune Dec. 14th, 1910.

The oldest son of General P.G.T. Beauregard, René Toutant was born September 19th, 1843 in Plaquemines Parish, LA.[1] His childhood home was "Magnolia," a sugar plantation in his mother's family, on the Mississippi River thirteen miles above New Orleans. His mother died when he was ten, and while his father was posted away from home, René was raised by his maternal grandparents and his uncle, Charles Villeré.[2] He was schooled at the future Louisiana State University, and declined an appointment there as commandant and professor of French.[3] When war began he joined as his father's secretary after the battle at Manassas, VA, with the Confederate rank of cadet from Louisiana. He wished to learn the discipline of artillery, and in Virginia in the summer of 1861 attached himself to a company of the Washington Artillery of New Orleans. He then allied himself with Col. William Ransom Calhoun's 1st SC Artillery, and returned to Charleston with the unit. Beauregard had several connections which made his subsequent appointment to Ferguson's Artillery Company a natural fit: Col. Calhoun's younger brother was a lieutenant in Ferguson's, and the Fergusons of Charleston likely knew the Beauregards of New Orleans.

From January to March of 1862 Beauregard recruited successfully for the company in Charleston and New Orleans. Not until he reached the age of twenty-one could he be commissioned as a lieutenant in the Confederate Army, despite the efforts of influential family friends to hasten the process, but he was a de facto lieutenant by the time Ferguson's left for Jackson, MS in the spring of 1863.[4]

Beauregard rose to command when Ferguson was wounded at Jackson and left field service, and the unit was his for the remainder of the war. The unit is known as Beauregard's Battery as well as Ferguson's Artillery Company in contemporary and post-war records. He was promoted to captain in April 1864, effective from December 21st, 1863. As the son of the influential Gen. Beauregard, René and his unit received special attention. The social ties between prominent Southerners of the officer cadre, specifically the Beauregards and the Fergusons, were evident from the beginning and no doubt facilitated the formation and supply of the company. Ferguson and 1st Lt. Beauregard were not shy about enlisting the general's patronage.

The general brought his son's unit to Mississippi with him for its first field assignment, and he brought it from the western theater to join him in NC for its last hurrah. After the company ceased as a fighting unit at Salisbury, NC in April 1865, Gen. Beauregard endeavored to secure last minute promotions to major for René and his other aides de camp, but President Davis declined to advance René. Beauregard was called a major after the war, but there is no official documentation that the rank was awarded. Capt. Beauregard was paroled as a member of his father's staff at Greensboro, NC on May 1st, 1865.

Beauregard was an attorney and judge in Louisiana after the war, and active in politics and Confederate Veterans' groups. In 1878 he married Miss Alice Cenas. He died December 14th, 1910, survived by his wife and six children, and is buried in the Metairie Cemetery, New Orleans, LA.[5]

Lt. John Ashe Alston

John Ashe Alston, Jr was born August 21st, 1842 at Georgetown, SC. His was a family of significant privilege. His grandfather was known as "King Billy" due to the family's wealth as one of the largest rice planters in the Georgetown area.[6] By 1860 Alston was pursuing a course of study in Europe, intent on becoming a physician.[7] When war came to SC, he secured a lieutenant's office serving under Thomas Ferguson, no doubt a family friend made in Charleston social circles. He was appointed in February 1862 and was recruiting soldiers into the unit that month.

The muster rolls suggest that 2nd Lt. Alston spent most of his time with the unit in the field, and much less time on recruiting duties. He did participate in efforts to return deserters to the unit, as when he was absent the first week of April 1863, leading Cpl. Carrier and Pvt. McCullough in the pursuit of miscreants. Alston was often left in command of the company when his superiors were away on other duties, listed as "commanding company" on his muster roll entries for requisitions. He was slightly wounded in the neck at Kennesaw Mountain.

He married Emma Rebecca Sanders, and they had three children.[8] Alston was a physician after the war, died of a brain tumor on May 6th, 1882, and is buried on Pawley's Island, SC.[9]

Lt. Henry Davis Calhoun

The Calhouns were some of the most prominent, wealthy and politically influential in nineteenth century South Carolina. Henry Davis was born into this rarefied family in 1838, to parents John Ewing and Martha Maria Davis. The family lived in Charleston, but escaped the summer heat in the Upcountry district of Pickens, where they had large land holdings. Henry was born there at their "Keowee" plantation, at the junction of the Twelve Mile and Keowee Rivers. John Ewing Colhoun favored the distinctive spelling of his last name with an "o" rather than the more familiar "a," but Henry Davis preferred the conventional "Calhoun."

His father died in 1847, and in the summer of 1850 young Henry was living at Keowee with his widowed mother and two siblings.[10] Before the onset of war, Henry and his younger brother, Edward B., were managing 2600 acres of the Calhoun family's interests.[11] The family owned eighty slaves, who jested that the father's wealth was so great his horses were shod with silver shoes.[12]

It would be expected that Charlestonians Thomas Ferguson and Henry Calhoun were, at the least, childhood acquaintances. It is therefore not surprising that Calhoun would obtain an office of lieutenant in Ferguson's company of SC artillery. As 1st Lt., Calhoun was heavily involved in recruiting, and made frequent successful forays into the SC Upcountry, where his family's name and his familiarity with the locals made it fertile ground for his enlistment efforts.

Henry's older and younger brothers were also artillery officers in 1862 Charleston. The oldest of the three, West Point graduate Col. William Ransom Calhoun of the 1st SC Artillery, under whom René Beauregard served before joining Ferguson's Artillery, was killed in September in arguably the most famous duel in SC history. The youngest, Capt. Edward B. Colhoun (the only one of the three sons who favored their father's spelling of the surname) served to the end of the war with Lucas's Battalion of Heavy Artillery.

After the wounding of Capt. Ferguson in 1863, Lt. Calhoun took on more of the day-to-day operation of the company, reflected by his requisitions of equipment and supplies. His activity in Columbia in the fall of 1863 suggests that he was raising another company of Ferguson's Artillery, whose men he likely accompanied to Atlanta. The last muster roll entries note his purchase of pistol cartridges in early 1865, in anticipation of what would turn out to be the unit's last action during the Carolinas Campaign.

Henry returned to his Keowee plantation after the war. "Captain Hal Calhoun" was one of the mounted participants in a theatrical production in October 1865.[13] His nickname might also have been "Hall."[14] There is no record of his promotion to a captaincy, but wartime references to him as "Captain" suggest that he was at least acting as one while recruiting a possible second company of Ferguson's Artillery. Calhoun died in Pickens, SC on March 27th, 1866; no obituary appeared in the local newspaper. He was buried next to his brother, William Ransom, in the family graveyard at Keowee Heights Plantation, in the present Clemson University Experimental Forest.

Joseph Anatole Hincks.
https://www.ancestry.com/.

Joseph Anatole Hincks was born July 17th, 1841 into a prosperous New Orleans family; his father, John William Hincks, was a physician. His family would have been friends with that of Gen. P.G.T. Beauregard, so it is natural that he ended up as an officer in René Beauregard's artillery company.

Hincks enlisted as a 3rd sgt. in Co. B of the Orleans Guards, a militia unit which was transferred into the service of the Confederate States. He was at the battle of Shiloh and his "health failed him under the hardships of the battlefield and the bad conditions at Corinth." Furloughed to New Orleans on the advice of Gen. Beauregard, he "narrowly escaped" when the Federals captured the city (one record said that he was simply exchanged). He described his struggle through the swamps and canebrakes, crossing Lake Pontchartrain in a cat-rigged boat, enduring a terrible storm at Biloxi, MS, and dodging Federal gunboats, to make his way to Mobile, AL where he enlisted in Ferguson's Artillery Feb. 4th, 1862.[15] He found the "company composed of such material as to render it incompatible with his social position." This is further evidence of the wide social gulf between the officer corps and the private soldiers. Benefiting from his family friendship with Gen. Beauregard, Hincks preferred to join him as a clerk at the HQ of the Department of SC, GA, and FL on detached service from Ferguson's. Hincks mustered there from Sept 1862 until June 20th, 1863 when he was released from that extra duty billet to report to Lt. Calhoun in Ferguson's Company.

When a lieutenancy in FAC opened, two men were recommended for the office. The choice was between Hincks and 'Iön Simons. Ferguson preferred Simons somewhat, but was not opposed to Hincks, and even wrote a letter recommending his appointment on Dec. 16th, 1863, well after Ferguson's wounding and cessation of field command. Since Hincks had been in the company as an acting officer, even though ranked and paid as a private for the better part of a year, the brass felt that he would be the best choice, avoiding a possible deleterious effect on "esprit du corps" from appointing Simons, who was an outsider to the men of the company. Hincks had left his clerk's position at HQ with the expectation that Ferguson's was to spawn a Company B, forming a battalion of artillery, and therefore accommodating him as an officer. (Hinks was at Atlanta with a detachment of the company from October through December of 1863; it is likely that those forty-some men and an equal number of horses were meant to be the nucleus of the new company). That plan failed to come to fruition, and a frustrated Hincks, who had been serving since June 20th as an officer without its pay or perks, wrote a letter the following March from Dalton, GA in which he declined the forthcoming appointment as lieutenant, deferring to Simons. He requested and received a transfer back to his former duties as a clerk at Charleston, to escape the insupportable burden of further service as an enlisted man with the private soldiers of Ferguson's Artillery. He enlisted May 7th, 1864 in Co. D, Manigault's Battalion of Artillery, and continued as a clerk on detached service at Department HQ until war's end.

Hincks returned to New Orleans after the war and died there Dec 5th, 1923. He is buried in the Metairie Cemetery in New Orleans.[16]

'Iön Simons

Simons was born Sept 3rd, 1839 in Charleston, one of many children of a cotton factor. The family was mistakenly listed as Irish in the 1850 census, corrected in later censuses.[17] His unusual first name was spelled in various ways by other family members who carried it: 'Ion; I'on; Ion; and 'Iön (this is how he signed a form in the muster rolls for Ferguson's Artillery). Coming from a wealthy family, 'Iön was well-educated, a clerk, and destined to be an officer when the war came.

Simons joined his brother, Samuel Wragg, and enlisted March 25th, 1862 as a twenty-two year-old private in the Charleston Light Dragoons, later called Co K of the 4th SC CAV. Two other brothers also served in the Confederate army, one dying in service. (Samuel Wragg was accidentally shot in the neck in camp, and survived the war.) 'Iön was promoted to 4th corporal and served until June of 1863, when he transferred to Ferguson's Company as an acting 2nd lieutenant, without rank, in favor of private Hincks, to replace the retiring Lt. Stark.

Major Martin had recommended Simons as active and conscientious in the discharge of his duties as adjutant of Martin's Battalion of artillery, to which Ferguson's was attached. Simons was immediately ordered by Lt. Calhoun to begin recruiting for the unit, and traveled to the Upcountry of SC. His appointment as lieutenant was confirmed December 8th, 1864.

Simons went to California after the war, and tried his hand as a land agent and sheep farmer.[18] He died of a stroke in Charleston, a single man, on January 26th, 1899, and is buried in Charleston's Magnolia Cemetery, with brother Samuel Wragg and others of his family.[19] Ironically, Samuel Wragg named a son after his brother; the namesake, Ion, became Charleston's city electrician.

Louis Raoul Stark

Stark was born on June 4th, 1841 in South Carolina, where his father was a physician.[20] Louis was educated at the Citadel in Charleston, where he appears as an eighteen year-old student in the 1860 ward 5 census, with T.B. Ferguson, his future captain. Stark enlisted as a private in the Marion Artillery and two weeks later was appointed as a lieutenant in Ferguson's unit. He was assigned to detached duty with the Conscription Bureau as an enrolling officer in the Laurens District, and then returned to the unit late in 1862. His recruiting trip to Georgia and the Upcountry of SC brought three soldiers into the company in the spring of 1863. Stark was with Ferguson's for its campaign at Jackson, MS in the summer of 1863, and was present through the battle at Missionary Ridge that November. He resigned Jan. 4th, 1864 to become adjutant of the 10th SC INF, was wounded at the battle of Franklin in December, and served to the end of the war with the 10th INF.

After the war, Stark followed his father into Medicine, graduating from the New Orleans School of Medicine in 1867. He married Mary Cannon in 1873 and started a family before moving in 1878 to Arkansas, where he was an allopathic physician and professor at the University of Arkansas, and a Master Mason.[21] He died there of erysipelas April 9th, 1909 and is buried in Mount Holly Cemetery in Little Rock.[22]

Dr. William Izard Bull, Jr.

Bull was born Jan. 19th, 1838 into a wealthy slave-holding family in Charleston.[23] He graduated from the SC Medical College in 1860.[24] Bull enlisted as an assistant surgeon July 27th, 1861 in McCreary's 1st SC INF. When the Confederacy assimilated state units the next spring, standards were set for surgeons, and any already in service had to stand for an examining board to assess their competence.[25] Bull joined the Marion Artillery as a private on April 28th, 1862, from which he was detached for service as an operator in the Signal Corps. He passed the Army Board of Medical Examiners as an assistant surgeon on March 21st, 1863, and was posted to the Army of Tennessee, where he joined Martin's Battalion of artillery, which included Ferguson's company. Dr. Bull was with Martin's Battalion through April 30th, 1864, and when he surrendered at Citronelle, Al on May 4th, 1865, he was attached to the 5th Company Washington Artillery.

Dr. Bull plied his medical trade in Charleston for several years after the war, and then took sales and clerical positions with a fertilizer firm until he passed away July 29th, 1917. He is buried in Charleston's Magnolia Cemetery.[26]

3. Recruiting the Company

Ferguson's first recruits were enlisted in January of 1862 into South Carolina's regular army. The company was transferred to the Confederate States on April 4th. As of March 6th, 1861, regulations stated that a battery of artillery was to be six guns, and should include a captain, two first lieutenants, one second lieutenant, four sergeants, four corporals, two musicians and seventy privates.[1] Practical considerations led to Confederate batteries of only four guns, so eventual manpower requirements would have been less than first prescribed. On May 27th, 1862 Ferguson asked of command: "Will my company be filled to one hundred and fifty by conscription? My company is now equipped as a Battery of mounted artillery and numbers eighty-one (81) enlisted men." This inquiry was in response to a new law which increased the manpower in a battery of artillery.[2] By the end of the year, 115 men had been enlisted in the unit. The officers recruited aggressively in the spring of the next year, anticipating the need for replacements when they eventually entered field service, and to replace the losses of 1862 to disease and desertion. When Ferguson's left for the Jackson campaign in the spring of 1863, there had been one hundred eighty-seven men enrolled since recruiting began, some of whom had been lost, but certainly the unit was well-manned and at full strength.

As stated by Gen. States Rights Gist in a letter to the Secretary of War, Ferguson's "unyielding energy" had recruited so successfully that he had almost enough for two companies. In fact, at one point there were plans to split the unit into two batteries, but they never came to fruition. Private James A. James wrote his parents Aug. 20th, 1863: "I have good officers here and if Griffin James wants to transfer I can get him one here. I hear there is firing every day at Charleston and they are at it now, will try and get a furlough before I leave here but want another box...Camp near Columbia, Captain Calhoun's Light Artillery."[3] On Sept. 12th, 1863, Major C.D. Melton reported that "Ferguson's Battery has also sent out the nucleus of a new company under Captain Calhoun, which is now recruiting and fitting up as company No. 2 of Ferguson's Battery."[4] Horses were being provisioned in Columbia, SC by Calhoun and Lt. Hincks in September and October of 1863 for this new company. Hincks was in command of a detachment of Ferguson's in Atlanta from October through December, which rejoined the rest of the company at Dalton, GA in January of 1864, as plans for the new company had never matured.

Over half of the recruits came from the Upcountry of SC and northeast GA, chiefly due to the efforts of Lt. Calhoun, whose prominent family was well known there (Fig. 1). Fully thirty-five percent of the men came from Calhoun's home district of Pickens, SC, and another twenty percent from nearby SC and GA locations. New Orleans was a fertile recruiting ground in the spring of 1862, providing eighteen percent of the unit's men. Capt. Ferguson's apparent modus operandi was to enroll any male he could, regardless of the recruit's likelihood to successfully serve, as it was not rare for these New Orleans men to wash out early for reasons of age and disability. The enlistment bounty of fifty dollars was no doubt an incentive to potential recruits (Fig. 2) and Ferguson was willing to take a chance on suspect candidates in order to fill the company. His home in the coastal area of Charleston was the source for nine percent of enlistees, and the remainder of the men came from around the state and beyond.

None of the men of Ferguson's could expect to get rich in military service. The pay scale for Confederate officers in the light artillery was established March 6th, 1861. They were paid at the same monthly rate as for cavalry officers: major $162; captain $140; 1st lieutenant $100; and 2nd lieutenant $90.[5] Fuel and travel expenses were reimbursed, the last at ten cents per mile. Each of their private horses, the number of which was specified by rank (three for captains and two for lieutenants) was provided forage[6] or reimbursed at a rate of $8 per month if no feed was available. Officers had to provide their own uniforms and private arms.

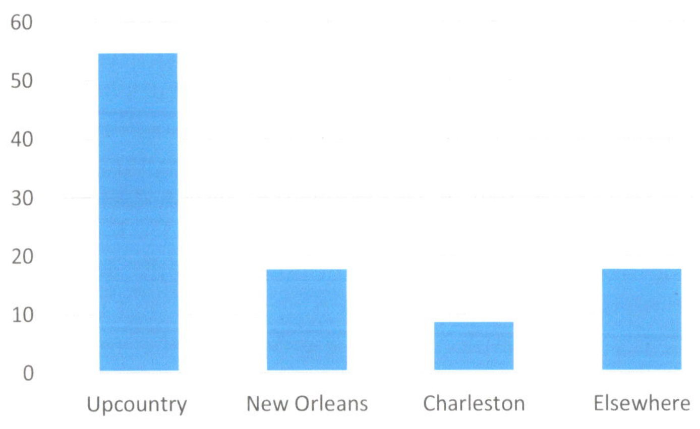

Fig. 1 Enlistment by Location.

Noncommissioned officers received cavalry pay, as well: 1st sergeants \$20; "buck" sergeants \$17; and corporals \$13. Privates in the light artillery earned \$12.[7] When Ferguson's men were paid at the infantry scale in November and December of 1862, the mistake was rectified the next pay period and they received two dollars more.

Additional duties merited higher pay, so that privates who were farriers, blacksmiths, articifers and musicians received the same pay as corporals. Extra duty pay of up to two dollars a day was available when a soldier was detailed from his regular field service to perform a task characterized by "uncommon skill or industry." A daily ration was to be provided for each enlisted man, and he was to be clothed at government expense. In times of need, when there were no supplies to issue, soldiers were paid by the "commutation system" to reimburse them for what they had spent on food and clothes.

It was not uncommon for the men of Ferguson's to go months without pay, particularly when they were on campaign in the field. They would receive back pay when they were on leave behind the lines, or in a hospital, or otherwise away from the front. Food rations could also be scarce, and depended on the stability of supply lines. Clothing wore out quickly, and was replaced when available. Issues of clothing to the unit are often the only records to establish the presence of a soldier during the last years of the war.

The men recruited into the company reflected the southern society from which they came. The officers were uniformly wealthy, educated, usually from a slave-owning family, familiar with an urban environment, and accustomed to a position of command and authority. Those in the rank and file were quite different. They were mostly rural and illiterate farmers, provincial, and were self-reliant, independent souls who sometimes chafed under the harness of military discipline. Many of them were European immigrants, and more were but first-generation Americans. Most of the men were white, but Ferguson had at least two persons of color in the ranks.

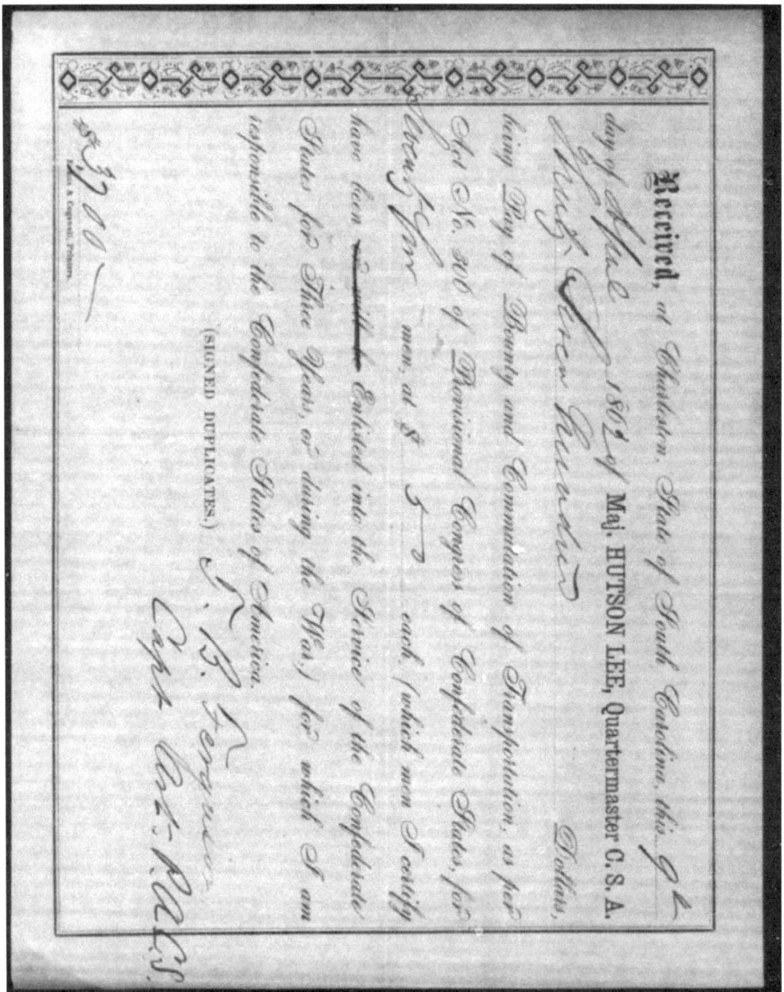

Received, at Charleston, State of South Carolina, this 9th day of April 1862 of Maj. HUTSON LEE, Quartermaster C. S. A. Thirty Seven Hundred Dollars, being Pay of Bounty and Commutation of Transportation, as per Act No. 300 of Provisional Congress of Confederate States, for Seventy Four men, at $ 50 each, (which men I certify have been ——— Enlisted into the Service of the Confederate States for Three Years, or during the War,) for which I am responsible to the Confederate States of America.

(SIGNED DUPLICATES.) J. B. Ferguson
Capt Cat. R.R.C.S.

$3,700

Evans & Cogswell, Printers.

Fig. 2 Receipt for Bounties.

Pvt. William Sizemore was enlisted March 25th, 1862 at Pickens, SC by Lt. Calhoun, very early in the formation of Ferguson's Artillery Co. The muster rolls show that he deserted after four months, was returned to the unit and served while awaiting a court martial in January. He was found guilty but the sentence was remitted in February, and he continued in service until captured on the retreat from Missionary Ridge. After ten months imprisonment at Rock Island, IL, he obtained his freedom by enlisting October 6th, 1864 for one year in the 2nd US Volunteer INF, Company D, for service on the western frontier. Forty years old, born in Pickens, he had dark complexion, eyes and hair. He was discharged from service at Fort Leavenworth, Kansas on November 1st, 1865. Claiming a hernia suffered in a fall from his military mount, partial paralysis, advancing age and fading strength, he received a US pension, and got his final $12 monthly payment on November 4th, 1904, just before his death on January 15th, 1905.[8]

There is nothing particularly unusual about Sizemore's military history, all aspects of it being shared by other soldiers of the unit. However, the extended Sizemore family in Pickens was remarkable for including whites, blacks, and possibly Indians, intermarrying and leaving a credible historical record that William Sizemore was not white. His great-grandson, Samuel Lorenzo Malone, wrote a family history which claimed that Sizemore was a white Irishman with a black common-law wife, and pointed out the irony of him fighting for the cause of slavery with a house full of his mulatto children.[9] None of the family members he interviewed for the history ever saw William. It was only family lore that he was white, and the only evidence Malone offers is reference to a twenty-year-old "William Sizemore" in the 1860 census of Pickens, SC, who is not to be found in an index or individual search of all families in the census.[10] This phantom Sizemore clearly could not be his ancestor, based on later census data which Malone provides but curiously never attempts to reconcile with the family's claim that their ancestor was white. Malone's research is suspect in other regards. He mistakenly claims that his ancestor served in Co. K of the 5th SC INF. In fact, that was a white man of the same name, born and enlisted in Spartanburg District rather than

Pickens, and some twenty years younger than the man who served in Ferguson's Artillery.[11] The 1850 census of Pickens, Eastern Div., family #662, shows William's mulatto mother, Sarah, and his black wife, Martha. William is not listed. None of the family can be found in the 1860 census. In the 1870 Pickens census, family #174, William, age 48, his wife, mother, and seven children, are all mulatto. In the 1880 census of Pickens C.H., family #262, William, age 54, is black, and his mother, wife and four children are mulatto. In 1900 Pickens, Distr. 121, family #254, William, born January 1827, age 73, his wife, and his son are all black.

When William applied for his US pension in 1892, he was described as 6'1" tall; dark complexion; gray hair; and dark eyes.[12] No racial categorization appears on the application. He claimed to be unmarried yet the censuses repeatedly list his "wife." Malone states that William and Martha never officially married[13] so William was likely adhering to the letter of the law when he admitted on his application form that he had no wife of legal record.

In four censuses over fifty years, not once were William Sizemore or his mother listed as white. One descendant's theory that he was a "Black Indian" was not proven by any documents available to his family historian.[14] All evidence points to William's initial allegiance as a Black Confederate. The irony was not that a white man with black children fought to preserve slavery, as claimed by Malone, but rather that a black man was fighting in the Confederate army, for reasons known only to him.

Another soldier of color in Ferguson's Artillery was Private William Dangerfield, a free mulatto blacksmith enlisted by Capt. Ferguson on April 15th, 1862 at their home parish of St. Johns Berkeley, Charleston, SC. (Fig 3). A white smithy would have been exempt from compulsory service due to this valued trade.[15] When Dangerfield decided to volunteer, his personal familiarity with its captain prompted him to enter the company led by T.B. Ferguson, whose family lived very close to his.[16] William was of white, black and perhaps Native American ancestry.[17] Dangerfield's mother, a freed slave, was multiracial, black, white and Native American, in varying proportions depending upon independent DNA tests on two living descendants.[18]

Dangerfield deserted Ferguson's while in Charleston on Sept. 7th, 1862, in order to join two older brothers in Co. E of the 5th SC CAV. A muster roll notation of "desertion" often referred to this offense: "No non-commissioned officer or soldier shall enlist himself in any other regiment, troop or company, without a regular discharge from the regiment, troop or company in which he last served, on the penalty of being reputed a deserter, and suffering accordingly."[19] He went to Virginia with his cavalrymen brothers, and when he obtained a lengthy sick furlough returned to Charleston, where he was paid salary due him for his six months service in Ferguson's company. He returned to the 5th SC CAV at least until Dec. 1st, 1864, the date of a clothing disbursement and the last of his entries in the military records.

Fig. 3

4. Camp Life: Crime and Punishment

Jan the 15th 1863

State of SC Sullivans Island

Dear sister I for the first time seat myself this evening to drop you a few lines in order to let you no that I am well at this time hoping that when these few lines come to hand they may find you injoying the same like blessing. I have fatened so since I come here that you would not no me I weigh about one hundred and sixty five or seventy pounds If I live to get to come I will surprise some of the girls up there you may tell them they may look out for if I live I am a coming tell Robert to be a good boy and tell them to be kind to each other and give mi best respects to all inquiring friends and especily to the one I love best Elisa Jane. You have them holes open in your ears for I am going to bring them fine ear bobs when I come home I saw Joseph Chastine the other day and he was well he is here on the island in mi old regiment. So no more at present time but remain your Brother until Death rite soon - Direct your letter to Sulivan Island Furgeson Lite Battery

William Chastine[1]

Private Chastain had been in the unit for eight months, after leaving his home in the SC Upcountry. Like many of his comrades, he was likely seeing the coast for the first time, and there would have been many lifestyle changes dictated by his new Army life. Purchasing earrings for his sister was but one of the many opportunities afforded by the bustling city of Charleston. Sullivan's is a barrier island off Mount Pleasant, which is across the Cooper River from the city (Fig. 1). The unit occupied camps near Charleston from its inception until it left for the Jackson campaign in the spring of 1863. Ferguson's was at a camp "near Charleston" from April to July of 1862, then at Camp Beauregard in Summerville until August, then on Sullivan's Island to the end of the year, and at Camp Helen in Christ Church, near Mt. Pleasant, until they left for Jackson. Its proximity to Charleston meant that the unit benefited from ready supplies during its formative months, but the downside was attractions which could prove detrimental to the discipline of newly-minted soldiers. Temptations abounded, and the muster rolls attest to the frequency of imprisonments for "absent without leave" and "desertion."

Fifty-four men in the muster rolls have notations of either AWOL, desertion, arrest, court martial, imprisonment, or a combination of these. This is twenty percent of the names, and the figure was likely higher, since records are incomplete for the later years. Most AWOL infractions were apparently considered minor since the perpetrator usually reappeared on active duty without any judicial proceedings. "Any non-commissioned officer or soldier who shall, without leave from his commanding officer, absent himself from his troop, company or detachment, shall, upon being convicted thereof, be punished according to the nature of his offense, at the discretion of a court-martial."[2] Being AWOL could be as innocent as overstaying a furlough due to difficulty obtaining return transportation, and seems to have been routinely excused.

Desertion would have been claimed if the soldier remained AWOL, and charges may have been preferred if he had to be captured and returned to the unit, but that crime also appears to have

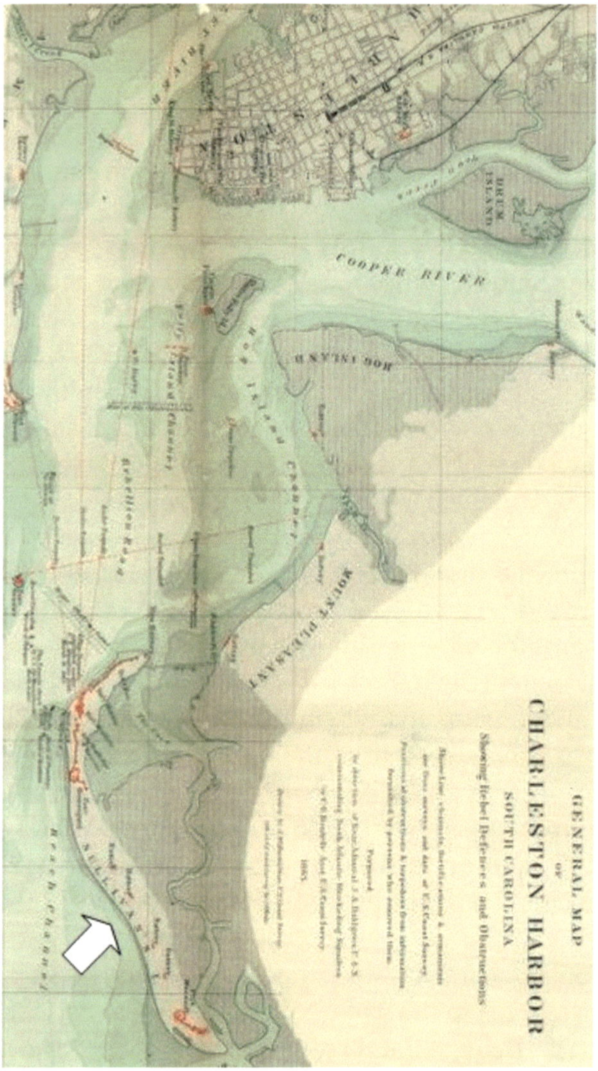

Fig. 1 Eugene Willenbacher, et al. *General Map of Charleston Harbor, South Carolina: showing Rebel defenses sic and obstructions* [Washington, D.C.: GPO, 1865]. https://www.loc.gov/0000205item/ Arrow: Sullivans Island.

been seldom brought to a court martial. The circumstances of the crime, the nature of the soldier, and the welfare of the company were factors in the punishments meted out by the officers. In the company's first eighteen months, Capt. Ferguson seems to have been plagued by disobedience in the ranks, desertion rankling him the most. It was taking some effort to instill a compliant attitude in his independent recruits. Ferguson clearly was a stickler for regulations, but only with flagrant desertion did he push for strict punishments. Desertion was a court martial offense and merited severe punishment: "All officers and soldiers who have received pay, or have been duly enlisted in the service of the Confederate States, and shall be convicted of deserting the same, shall suffer death, or such other punishment as, by sentence of a court-martial, shall be inflicted."[3] Even when a soldier was convicted at trial and sentenced to jail time, Ferguson often did what he could to free them, notably before the unit's first campaign when he wanted all of his trained soldiers present for duty.

On July 12th, 1862 Ferguson wrote to the asst. adjutant general: "I have the honor to communicate to Brig. Gen. Smith, cmndg this district that I have intercepted information of the whereabouts of four of my deserters. They have been writing to persuade some of the Company to join them, giving a plan of escape. I think I can apprehend them by taking immediate steps, so will leave for Atlanta where one of them is at present, via Augusta. I will then wish to proceed to Clarkesville where I can find the other three. The discipline of my command will be marred and I will not be able to keep it effectually together, unless these men are dealt with summarily..."

Desertion was not a crime to be ignored, but encouraging it was no less egregious: "Any officer or soldier who shall be convicted of having advised or persuaded any other officer or soldier to desert the service of the Confederate States, shall suffer death, or such punishment as shall be inflicted upon him by the sentence of a court-martial."[4] From July 15th to August 28th, Ferguson logged 1630 miles pursuing these deserters from Chattanooga and Knoxville, TN to Pickens and Anderson Districts in the upcountry of South Carolina.

The men were returned to face military justice, and they spent some months in prison awaiting their fates. On February 28th, 1863, Ferguson asked Gen. Jordan that three of the deserters, G.W. Rains, William Sizemore and Peter Burton, be released from arrest and returned to his company, they having been confined at hard labor while awaiting court martial decisions. The three were promptly released with their sentences remitted.

The fourth conspirator appears to have been Irishman Bernard Donnelly, likely the ringleader judging by his sentence.

Charleston Nov. 16th 1862

General Orders No. 9 from Brig. Gen. R.S. Ripley

> I. In accordance with General Orders No. 96 from Department Head Quarters bearing date Nov. 11th 1862 Private Bernard Donnelly of Capt. Ferguson's Company of Artillery P.A.C.S. will be "shot to death with Musketry" on the Front Beach of Sullivan's Island at 12 o'clock M. on Monday the 24th day of November 1862.
>
> II. All troops on Sullivan's Island will be turned out under command of Col. L.W. Keith Commd'g Sullivan's Island at 11 o'clock A.M. of the same day to witness the execution. The 30th Reg't SC Vols. Lt. Col. Dantzler Comd'g will be armed and equipped. All other troops will appear with side arms only.
>
> III. The Comd'g officer of Ferguson's Company will make a detail of twelve men from his Company, to carry out the sentence of the Court Martial, providing them with nine loaded and three unloaded muskets for that purpose.
>
> IV. A Reserve detail of twelve men with guns similarly loaded will be made in case they are needed to complete the Execution.
>
> V. Capt. T.N. Huquenin is charged with the details of, and execution of this order as Acting Military Provost Marshal.

Thus, an example was to be made of the hapless Irishman, but the execution never happened, and the record is silent as to the identity of his guardian angel and the reason for his salvation. He was back on duty by the end of December, leaving his three still-incarcerated accomplices convinced that "military justice" must be an oxymoron.

One misfit stands out in the officers' constant efforts to maintain military discipline. George Griffiths was twenty-four when he enlisted January 30th, 1861 at Charleston as a private for one year in Company B of the 1st Battalion Regular Artillery of the South Carolina Army. This unit would be transferred into Confederate service in May, and be renamed the 1st Regiment SC Artillery. Griffiths stood 5'7", had a dark complexion and hair, and black eyes. He got off on the wrong foot for unknown reasons, and missed muster in April, being in prison at Fort Moultrie, across the harbor from his comrades at Sumter. It isn't known if he was in jail during the April bombardment of Sumter, which fell on the 13th. He was released and present at the unit's May 24th muster at the surrendered fort. Army life did not agree with him, somehow, and on October 20th he deserted. The circumstances of his desertion and subsequent punishment are not in the record, but he likely served more time in jail. At the end of his twelve months in the 1st Artillery, Griffiths opted for a change, and on February 6th, 1862 he was enlisted at Charleston by Lt. Alston as a private in the newly organized Ferguson's Artillery Company for three years or the war.

Turning over a new leaf didn't work out for him, and by June 23rd he was in Charleston's City Guard House for unknown offenses. He was back on duty by the end of August, but temptation was overpowering, and on October 20th he deserted again. He was returned Nov. 2nd to jail in Charleston, his fourth incarceration since becoming a soldier. He was listed present in camp at the end of February, an error as indicated by a letter he wrote from prison March 4th. (punctuation added for clarity)

> Resp. Sir
>
> Wishing to be transferred from Captain T.B. Ferguson's company to Captain Watie's now stationed at New Bridge. I should feel ever grateful to you sir if you would remit my sentence. I got three months by a General court martial and I

have put one month and a half of that time in. my sentence is the only thing that keeps me from transferring. I have received no pay for this last eight months and I have got a wife and family to support and it has gone very hard with them since I have been in this company only living by the support of making uniforms and sir I should be ever grateful to you if you would remit my sentence. I should very much like to go to Captain Watie's company and I believe sir it is my sentence that is keeping me from going to Captain Watie's. Both of the Captains are satisfied to transfer and I believe the papers are all ready made out and the person who is to transfer with me is Mr. Murphy.

I remain very respectfully yours
G. Griffiths

He had been found guilty in December of being AWOL and was sentenced to three months wearing a ball and chain. Dennis Murphy had originally been in John Waties' Co. B, Palmetto Battalion Artillery, and deserted to apparently join Ferguson's company in July 1862, earning himself a jail sentence. Waties had returned Murphy out of jail into his company on Feb. 20th, and was willing to accept Griffiths in exchange for returning Murphy to Ferguson's, but Ferguson could not abide the proposed arrangement. On March 12th he wrote from Camp Helen; "Since his enlistment Privt. Griffith has given me more trouble than any other man in my Co. I would therefore urge that his sentence be carried out. I think that it will have an injurious effect for him to be transferred. He is the only man in the Comp. who can blow the "Bugle Calls." I cannot therefore consent to his being transferred."

Griffiths knew how to work the system, and appealed to a higher authority. He wrote to Gen. P.G.T. Beauregard, his lieutenant's father, on May 5th, stating that his sentence had expired on April 17th, and that he was still in jail, without clothing and blankets. He requested transfer to "any company or even a gun boat as it seems impossible for me to get along although I do all I can." The obsequious Griffiths said he would be pleased to serve under Lt.

Beauregard if he had a command, evidently meaning to undermine Ferguson and have him demoted in favor of the general's son. Griffiths failed to inform the general that he was then in solitary confinement because he had been properly released on April 7th, and on that day violated the term that he be confined to camp; he foolishly ventured out for a better view of the attack by the Federal fleet in Charleston Harbor. He was released from jail on May 25th, after most of FAC had left for Jackson, MS, and asked for a transfer to an infantry company. That didn't happen and Ferguson got his bugler back, until Griffiths was captured May 18th, 1864 at Calhoun, GA. He saw his final jail sentence of three months at Camp Morton, IN, where he took the oath of allegiance.

5. The Weapon

Captain Ferguson had to go to some lengths to obtain cannon for his company. His Apr. 19th, 1862 answer to a telegram from Gen. Beauregard shows the nature of his relationship with the powerful general and hints at the patronage which benefited the company throughout the war.

> Dear General, Your telegram of the 14th inst. to René offering us a battery in your Army has been received and I hope is now answered by him. I take this my earliest opportunity to thank you for this additional mark of your kindness to me – I trust General that before the War is over we will have proved ourselves worthy of your favor.
>
> I can in a very short time procure the horses in this State if the authority and means for purchasing them be given me by the War Department, and there will be no difficulty in getting harness if orders be given for their issue at once, as there are several sets now in the Ordnance Department in Charleston.
>
> My Company has been turned over to the Confederate States and received as "Light Artillery" and is unequipped and unattached to any Corps so might very easily be ordered to your army.

I have attached to my command Capt. S.D. Lee's Company of Confederate Regulars under Lieut. T.B. Hayne C.S.A. it having been reduced by desertions and other causes to fifty-three.

I would be exceedingly obliged to you General, if you would apply to the War Department for my company and above mentioned detachment to be ordered to your command and for authority for me to purchase and bring on the necessary horses.

I see no prospect of our being furnished a Battery in this department for some time. I am now on my way to Columbus Geo. where I have been ordered by Genl. Pemberton to try and purchase a Battery But I think it very doubtful that I succeed. I left René in Charleston. He came up from Coles Island where we are at present located to meet his uncle – he is very well but exceedingly anxious to be in the field with a section.

Do General remember me to Brother & Drs. Brodie & Chappin. Hoping yet to be able to join you again

I am with much Respect Your obdt Servt T.B. Ferguson

Ferguson's doubts about available cannon were confirmed when he returned from his mission to Columbus empty-handed a week later. Gen. Beauregard did what Pemberton couldn't, and on May 2nd Capt. Ferguson was ordered to travel from Charleston to Columbus, MS to secure a four gun battery left there for him by Beauregard's Special Order No. 313: "Capt. T.B. Ferguson will proceed to Columbus Mississippi and take charge of a Battery of four pieces of artillery with harness left there for him by order of Genl Beauregard, and return with the same without delay to Charleston." (Fig. 1) Ferguson returned from there on May 25th, presumably with the cannon, type unknown. These guns were the ones used by the company for the next year in Charleston to learn and practice gun drill. Practice firing with live rounds was performed in April, June and October, and blank cartridges for practice were issued for purposes of drill.[1]

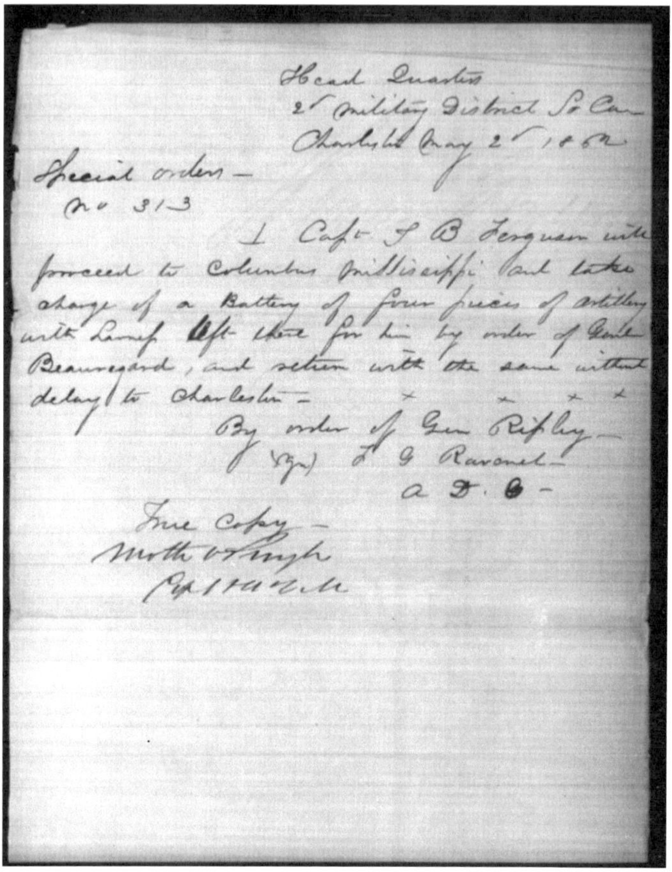

Head Quarters
2d Military District So Ca
Charleston May 2d 1862

Special order —
No 313

1 Capt T B Ferguson will proceed to Columbus Mississippi and take charge of a Battery of four pieces of artillery with Limbers left there for him by order of Genl Beauregard, and return with the same without delay to Charleston — x x x x

By order of Gen Ripley —
(Sgd) J G Ravenel —
A D C —

True Copy —

Fig.1. Special Order No. 313.

A year later the company was outfitted with four Napoleons received by Lt. Beauregard May 12th, 1863 at Montgomery, AL on the way to its baptism of fire at Jackson, MS (Fig. 2). These guns were provided by Richard Lambert of the Macon (GA) Arsenal, and were presumably manufactured in that city (Fig. 3).

Ferguson's was favored to receive Napoleons, formally called "12 pounder guns, model of 1857." It is a smoothbore weapon, and was a workhorse in both Federal and Confederate armies. The tubes manufactured at the Macon foundry, and some other southern Napoleons, lacked the muzzle swell of the US style. Usually of bronze, as were Ferguson's, Napoleons have a 4.62 inch bore, and fired several types of ammunition. Most of the projectiles were spherical, hollow iron balls containing a bursting charge of black powder, with a time fuse set for the estimated time of flight to the target, inserted into the shell just before loading. Some of these balls were "shell," whose bursting charge fragmented the thick shell wall into deadly missiles. Some of the balls were "case shot" with thinner walls, and contained, in addition to the bursting charge, about eighty balls under an inch in diameter.[2] Projectiles were oriented in the bore with fuses forward, and strapped to a cloth bag containing between two and two and a half pounds of black powder, with an intervening wooden "sabot." This arrangement of powder bag, sabot and fused projectile was called "fixed" ammunition.

The third type of ammunition for a Napoleon was an antipersonnel round used at ranges within three hundred yards. This was called "canister" because it was a thin-walled metal cylinder containing twenty-seven iron balls each about one and a half inches in diameter and just under a half pound each.[3] The force of the cannon's charge broke open the thin canister as it left the muzzle, and the effect was similar to that of a giant shotgun as the balls continued down range to the target, usually massed infantry. In the face of enemy troops at very close range, two canister rounds could be fired at once, a devastating "double canister". Ferguson's received 156 rounds of shell, and 48 each of shot and canister for the two pairs of guns in its battery (each pair is called a "section"). Each ammunition box was designed to hold thirty-two rounds, and as each gun had access to four boxes, each piece could carry 128 rounds for immediate use.[4]

A bronze Napoleon barrel and the two-wheeled wooden carriage on which it was mounted weighed almost twenty-four hundred pounds. Each cannon used another two-wheeled cart called a limber to carry an ammunition box, and when it was time to move the gun, its carriage was hooked to the limber, making a four-wheeled horse-drawn vehicle. With a full ammunition chest, when "limbered up" and ready to move, the entirety (called the "piece") weighed just under two tons.

Fig. 2 Southern Napoleon, Gettysburg National Battlefield.

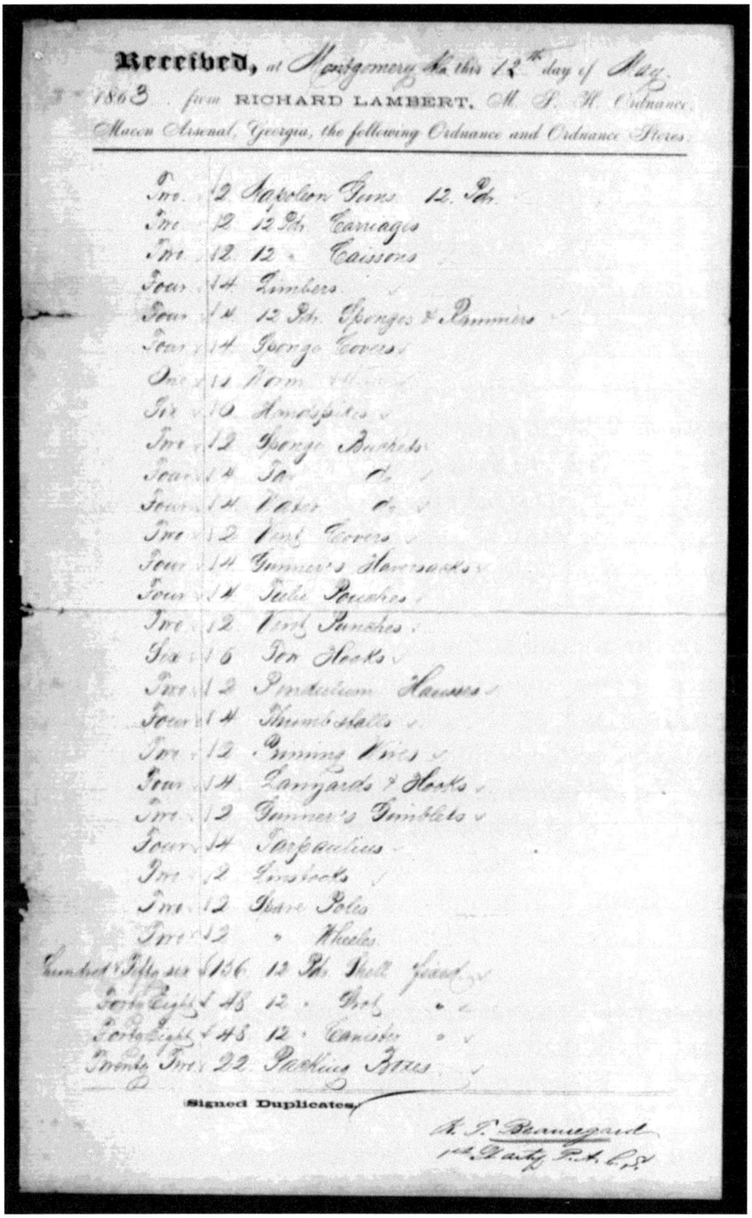

Received, at Montgomery Ala this 12th day of May 1863 from RICHARD LAMBERT, M. S. K. Ordnance, Macon Arsenal, Georgia, the following Ordnance and Ordnance Stores:

Two 2 Napoleon Guns 12 Pdr.
Two 2 12 Pdr. Carriages
Two 2 12 " Caissons
Four 4 Limbers
Four 4 12 Pdr. Sponges & Rammers
Four 4 Sponge Covers
One 1 Worm [illegible]
Six 6 Handspikes
Two 2 Sponge Buckets
Four 4 Tar do
Four 4 Water do
Two 2 Vent Covers
Four 4 Gunner's Haversacks
Four 4 Tube Pouches
Two 2 Vent Punches
Six 6 Pen Hooks
Two 2 Pendulum Hausses
Four 4 Thumbstalls
Two 2 Priming Wires
Four 4 Lanyards & Hooks
Two 2 Gunner's Gimblets
Four 4 Tarpaulins
Two 2 Linstocks
Two 2 Spare Poles
Two 2 " Wheels
One hundred & Fifty six 156 12 Pdr. Shell fixed
Forty Eight 48 12 " Shot
Forty Eight 48 12 " Canister
Twenty Two 22 Packing Boxes

Signed Duplicates.

R. T. Beauregard
1st Lt Arty P.A.C.S.

Fig. 3 Receipt for two of Ferguson's four Napoleons.

Ferguson's Artillery Company functioned as a unit of "mounted artillery." The term is a confusing misnomer, as only the drivers and the officers rode on horseback, the remainder of the men usually on foot, or riding on the limbers when speed was a necessity.[5] In the Confederate service, four horses were the motive power for the two ton load, so a battery of four Napoleons needed sixteen horses for the cannons alone. In addition, the battery had four caissons, carriages used to carry two extra ammunition chests and a spare wheel. Each caisson was hooked to a limber carrying another ammunition chest, and these four conveyances required another sixteen horses. A travelling forge for ironwork and horseshoes was needed for each battery, requiring four more horses. A battery wagon carrying additional supplies, an ambulance, ammunition wagon and wagons for forage were also pulled by draft animals. When the unit left for field duty, it would have taken sixty to seventy animals, and a prodigious amount of forage to keep them going.

The stresses of overwork, disease and inadequate nutrition were as hard on the animals as the men, and replacements were constantly required for both. When the Army of Tennessee (AOT) artillery was inspected in March of 1864, it was noted that "the animals were for the most part,. certainly thin; some had died from hard usage and disease not uncommon in our artillery service and others were worn down below the standard for use." This was felt to be chiefly unavoidable, and rationalized that the animals were seasoned and inured to hardship. "Extreme care will not always keep artillery horses in good condition, necessarily subject as they are at times to excessive draft, injurious exposure, and long fasting. The end of the fall campaign generally leaves them seriously reduced; the vicissitudes of winter and the difficulties in the way of supplies of forage sufficient in quantity and suitable in quality scarcely allow of their being adequately recruited by spring. In the meantime diseases will more or less prevail, and the animals are therefore often found still below a satisfactory standard in numbers and in flesh when a new campaign is about to open."[6]

Equine diseases were treated with various remedies. During the winter of 1863-64 at Dalton, GA, the animals were experiencing cases of the "scratches," the "sours" and vermin. "For scratches, wash clean and apply soft soap and grease; also bluestone, copperas, white-lead, poke-root and red oak bark have been used. For vermin a strong decoction of tobacco has proved efficacious, with sulfur given in the food." The lack of a veterinary surgeon was noted and acutely felt in the Army of Tennessee.[7]

Each of the guns was served by a crew of ten, using a drill which was taught and practiced until the soldiers were cross-trained in each of the positions at the gun.[8] Napoleons are muzzle-loaders, ammunition being loaded and rammed from the front (muzzle) to the back (breech) end, where the powder bag is punctured by a bronze priming wire introduced through the vent hole in the top of the barrel. This exposed two and a half pounds of propellant to the flame of a friction primer tube placed into the vent and ignited by a pull on the lanyard cord. Recoil of the gun required re- aiming before the next round was fired. This drill demanded attention to detail, cooperation and physical strength to avoid a potential catastrophe, most often a "premature ignition." This tragedy occurred when the crew failed to adequately employ the moist sponge-rammer to extinguish any lingering sparks in the barrel before loading the next round, which could explode while the men at the front of the gun were in harm's way. Even when not receiving enemy fire, the crew was in constant peril during firing.

When not on the battlefield, the men were busy with other duties. Fielding a battery of artillery requires the services of many skilled men. Since mounts and draft animals were integral to the function of the unit, men would have to groom the horses and mules, obtain their feed and water, maintain their leather harnesses, treat their diseases and injuries, and keep them shod. Blacksmiths were essential to provide shoeing and farrier services. Carpenter skills were needed to repair the carriages, which were subject to considerable stress and broke often. The list of raw materials and tools supplied to the battery (see next chapter) gives a sense of the skills required to keep it functioning, explaining why those with special talents were paid above the base wage for a mounted artilleryman.

Securing adequate feed for the animals was a constant struggle, and the officers of Ferguson's were frequently traveling, sometimes far from the company, to secure it. The prescribed daily ration was fourteen pounds of hay and twelve pounds of grain.[9] Following is a chronologic list of forage requisitioned by the named officers, and gives an idea of the fluctuating number of animals in service. There were no doubt additional unrecorded forage receipts. It cannot be assumed that any one requisition represents the total animals present on that date, since sometimes multiple disbursements were made in different places on the same day. There are several requisitions in the list for fuel, as well.

9 June '62 at Charleston, for 33 horses/12 mules; Capt. Ferguson
18 June '62 at Charleston, for 1 private horse; Ferguson
27 Nov. '62 for 1 private horse; Ferguson
1 April '63 at Camp Helen, SC for 4 private horses; Ferguson
1 May '63 at Mount Pleasant for 2 private horses; Lt. Beauregard
6 May '63 at Augusta, GA; Beauregard
10 June '63 at Charleston for 26 horses/8 mules; Lt. Alston
30 July '63 at Morton, MS; Beauregard
26 August '63 at Morton, MS; Beauregard
1 Sept. '63 at Pollard, AL for 37 horses; Beauregard
1 Sept. '63 at Kingston, GA; Beauregard
1 Sept. '63 at Columbia, SC for 12 horses; Lt. Calhoun
4 Sept. '63 at Atlanta, GA for 32 horses en route to Chattanooga; Beauregard
4 Sept. '63 at Columbia, SC for 28 horses; Lt. Hinks, commanding
17 Sept. '63 at Rome, GA; Beauregard
1 Oct. '63 at Chickamauga, TN; Beauregard
1 Oct. '63 at Columbia, SC for 41 horses/4 mules; Calhoun
6 Oct. '63 at Columbia, SC for 40 horses/4 mules; Calhoun
10 Oct. '63 at Chickamauga, TN for 74 horses/28 mules; Beauregard

21 Oct. '63 at Atlanta, 1 ½ cords wood fuel for 3 subalterns/37 privates and noncoms; Hinks, commanding

27 Oct. '63 at Atlanta, 5 bushels of charcoal for "detachment of Ferguson's," Hinks, commanding

3 Nov. '63 at Atlanta, for 40 horses/4 mules; Hinks, commanding detachment

21 Nov. '63 at Atlanta, for 40 horses/4 mules for the next five days; Hinks, commanding

27 Nov. '63 at Atlanta for 2 ½ cords wood fuel for 3 subalterns/40 privates and noncoms; Hinks, commanding detachment

30 Nov. '63 "in the field" forage; Beauregard

16 Dec. '63 at Atlanta, for 40 horses/4 mules; Hinks, commanding

21 Dec. '63 at Atlanta, for 40 horses/4 mules; Hinks, commanding

24 Dec. '63 at Atlanta, for 37 horses/4 mules; Hinks, commanding

Dec. '63 at Atlanta, for 2 ½ cords wood fuel for 3 subalterns/40 privates and noncoms; Hinks, commanding detachment

31 Jan. '64 near Dalton, GA for 74 horses/12 mules; Hinks, commanding

29 Feb. '64 near Dalton for 64 horses/14 mules; Hinks, commanding

30 Apr. '64 "army of TN" for 72 horses/13 mules; Beauregard

May '64 "army TN" for 67 horses; Beauregard

31 Jun '64 "army TN" for 55 horses; Alston

6. Supplying the Company

Building an artillery company from scratch would have been a monumental task. Over one hundred fifty men and animals had to be acquired, then housed, fed, clothed/shod, nursed, and replaced when needed. Capt. Ferguson and his officers would have spent much of their time assessing the needs of the company, filling out requisition forms, and haggling with the quartermaster department to obtain necessary equipment. The first year was spent in camps near Charleston, so supplies would have been easier to find, though never guaranteed. In the field on campaign, the supply lines were longer, and equipping the company was more problematic. An examination of the requisition forms in each officer's muster roll gives an idea of the wide variety and amounts of material which were needed to supply the battery. The following list is a chronologic summary of the items requested or supplied, exclusive of forage and fuel rations (see Chapter 5). The requested equipment was often not available and therefore not issued, but the requisitions give a good idea of the company's requirements.

50 blankets; 8 camp kettles; 20 mess pans, 5 Apr. '62, Charleston

60 pair shoes; 100 overcoats, 19 Apr. '62, Charleston

75 yards red cloth for collars and cuffs, "red is required by regulations," 26 Apr. '62, Charleston (Fig 1)[1]

9 Sibley, 3 officer, 1 commissary tent(s); 270 lbs. rope; 6 water buckets, axes, hatchets and spades; 12 curry combs and horse brushes; and 1 bay horse, 1 May '62, Charleston

80 artillery caps, 30 May '62, Charleston (Fig. 2)

20 lb. wagon axle grease, 4 June '62, Charleston

40 halters; 80 haversacks; and 1 packing box, 8 June '62, Charleston

50 pair shoes, 14 June '62, Charleston

50 lb. cut nails for repairing stables and quarters, 10 Dec. '62, Sullivan's Island, SC

400 ft. one-inch boards and 150 ft. 3x3 in. planks to repair quarters, 28 Dec. '62, Ft. Moultrie, SC

10 canteens, canteen stops and haversacks, 13 Jan. '63, Ft. Moultrie

16 Army tents; 12 axes; and 1 wall tent, 28 Jan. '63, Ft. Moultrie

1 coffin for Pvt. Joshua Spencer, 2 Feb. '63, Sullivan's Island

20 blankets; 30 pair shoes; and 15 pants, 4 Feb. '63, Charleston

One coffin for dead soldier, 11 Feb '63 (William Chastain d. 10 Feb.) Ft. Moultrie

5 Sibley and 2 wall tents returned as "perfectly worthless," 21 Feb. '63, Charleston

50 pair of shoes, 5 Mar. '63, Charleston

Requested 60 artillery jackets, blue pants, uniform caps, blankets and overcoats; 100 flannel shirts, drawers (underwear), socks; and 8 tents; received only 100 drawers; 20 blankets; and 8 tents, 24 Mar. '63, Charleston

2 bugles; and one 4 horse wagon, 26 Mar. '63, Charleston

4 sides (208 lbs.) leather; 4 gal. neatsfoot oil; and 2 gal. sperm oil, 5 May '63, Charleston Arsenal

4 12 pounder Napoleons and equipment, 12 May '63, Montgomery, AL

1 sorrel horse, 10 Jul. '63, Jackson, MS

1 coffin for soldier, "necessary from cause of disease," 20 Jul. '63, Ft. Moultrie

40 horses to complete section of artillery, 21 Jul. '63, Columbia

Pens; ink; pencils; paper; sealing wax; morning report forms; and two padlocks and curry combs, 31 July '63, Charleston

120 ft. rope; 2 prolongs; planes; files; saws; lanterns; packing boxes; halters; and brushes, 10 Aug. '63, Columbia[2]

4 pairs shoes, two jackets, 15 Aug. '63, Morton, MS

1 horse, 20 Aug. '63, Morton

30 jackets and pants; and 27 pair shoes, 22 Aug. '63, Morton

100 ft. rope and one wagon, 24 Aug. '63, Charleston

19 horses returned, "unserviceable," 25 Aug. 1863, Morton

1 horse; 14 mules; 3 wagons; 12 harnesses; and 3 saddles & stretchers, returned, "unfit," 25 Aug. '63

23 horses and 8 water buckets, 1 Sept. '63, Montgomery

20 shirts, drawers, blankets, jackets, pants, 7 Sept. '63, Charleston

20 pairs of shoes; and 3 A tents, "to complete section of Ferguson's Artillery" 9 Sept. '63, Columbia

4 tow mules and harness "to draw our baggage," for "my company ordered west" 28 Sept. '63, Columbia

3 camp kettles, 30 Sept. '63, Charleston

20 bushel bags; 100 ft. halter rope; 1 sack of salt, 4 lb. saltpeter; 1 gal. turpentine; and juniper oil (last four items for horse drench), 1 Oct. '63, Columbia

106 shirts; 60 pants and jackets; 87 pair socks; 15 hats; and 3 pair shoes, 7 Oct. '63

5 bushels charcoal, 4 axes and handles, 27 Oct. '63, Atlanta, GA

5 quires letter paper; 3 pen holders; 36 pens; 1 blank book; 1 bottle ink; 150 envelopes, 22 Nov. '63, Atlanta, Lt. Hinks, commanding a detachment of Ferguson's L.A.

8 blank books and 1 morning report book, 2 Dec. '63

1 English pistol belt and cap box, 9 Dec. '63, Charleston Arsenal to Lt. Beauregard

70 blankets, shoes, shirts, drawers, jackets, pants, caps and socks, 18 Dec. '63, Charleston

Supplying the Company

One morning report book, 8 blank books, 19 Dec. '63, Charleston, SC to Lt. Beauregard

14 horses returned unserviceable 31 Dec '63

48 drawers; 5 pair pants; and 15 shirts, 31 Dec. '63, camp in the field

15 horses; 6 mules; 4 horse wagon and harnesses; 30 jackets; 17 pants; 15 shoes; 7 shirts; 13 socks; 20 blankets; 8 mess pans; 1 coffin; 1 pick, shovel, axe and handle; 2 gal. neatsfoot oil and soap; 3 lb. tallow; 4 lb. nails; 5 lb. rope; 1 ½ bushels salt; and 16 horse and mule shoes, 29 Feb. '64, camp near Dalton, GA

4 pair shoes; 4 drawers; 10 lb. soap; knife; horse/mule shoes; nails; salt; axes; kettle; and rope, 31 Mar. '64, camp near Dalton

2 artillery horses; 10 tin cups; and 150 lb. salt, 30 Apr. '64, army TN

24 horses; 8 halters; and 150 lb. salt, 30 June '64, army of TN

10 artillery horses; 32 horse shoes; 2 lb. horse shoe nails; and 12 lb. salt, 24 Aug. '64, army of TN

3 horses, 31 Aug. '64, army TN

20 Navy cartridges, 21 Feb. '65, Macon Arsenal, GA

20 Army cartridges, 13 Mar. '65, Macon Arsenal

Fig. 1 Artillery enlisted uniforms.

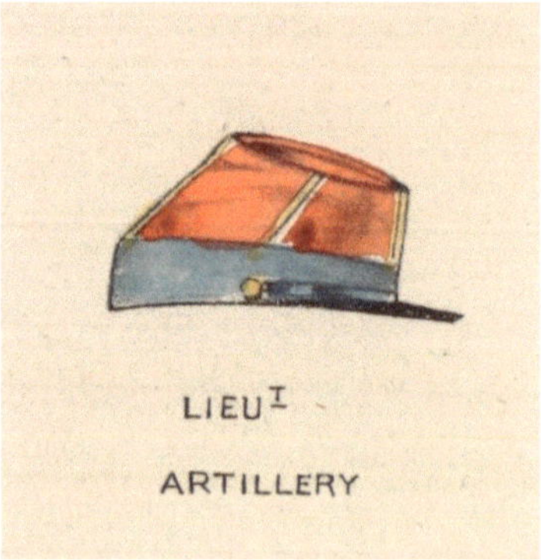

Fig. 2 Artillery Cap, 2nd style.

7. Losses: Casualties, Captures and Sickness

Forty soldiers died during the period for which muster rolls exist. Likely there were more late-war, unrecorded deaths. This is at least fifteen percent of the 273 men listed on the rolls. The causes were probably mostly disease, but the record is usually silent as to cause of death. Three men were specifically noted as being killed in action, and one died from wounds. Five died while prisoners of war.

Many more men were sick at some point during their service. For the men listed in the muster rolls, eighty-nine were noted to be on sick furlough, hospitalized, or sick in camp at some point. This represents thirty-three percent of the total enrollment, but the real numbers were certainly higher. Most of the soldiers were from rural areas, isolated and therefore never before exposed to the diseases which were omnipresent in the crowded conditions of camp. A recurring theme in a recruit's muster roll is an early hospitalization in the spring or summer of 1862, as the soldier first encountered typhoid fever, mumps, chicken pox, measles, and other pathogens, sickening and perhaps killing him from dysentery, fever or pneumonia. The sick call became more important during the campaign season, when unhealthy soldiers meant reduced unit effectiveness. The toll of reduced rations, increased exertion and unhealthy camp hygiene guaranteed that men would continue to fall out of the ranks as the war ground on.

Medical care at the time was rudimentary, made more so when what effective medicines as were known became scarce due to the blockade of southern ports. Hospitalization was no guarantee of recovery, as they were breeding grounds for communicable diseases. Surgery had limited application, and was attended with the risk of infection and death. Recently certified Assistant Surgeon William Izzard Bull, Jr. was present with Ferguson's by the time of the Jackson campaign, but his time was likely spread thinly with other artillery units in the battalion. He would have treated mild disease in camp, but more severe sickness and injuries would have been transported to a receiving hospital for transfer to a general hospital, if needed. The frequent notation of "died in camp" in the muster rolls suggests the rapid course of disease and the scarcity of effective medical resources.

The toll of disease is dramatically told by the tale of the Benefield family. Five brothers enlisted the same day in Ferguson's Artillery Company, and none returned home. The family left North Carolina and migrated to Heard County, Georgia, and by 1860 there were eleven children. The five oldest men were students, and unlike most soldiers in Ferguson's, were literate and part of a slave-holding family.[1] It isn't known why they didn't join the army earlier, but they were subject to conscription when they travelled to Atlanta, and on December 5th, 1863 were enlisted in Ferguson's Company by Capt. McClanahan, probably a conscription officer. The battery had lost men and guns on the retreat from Missionary Ridge in November, and were refitting with the Army of Tennessee in Dalton, Ga. The brothers would have taken the Western and Atlantic RR north from Atlanta to Dalton to muster in as privates for three years, or the war.

The brothers were: Henry, 33; John, 24; James B., 21; Joseph J., 19; and George W., 18. On December 28th George W. died in camp near Dalton, likely from an aggressive disease. The next day, Henry, James B. and Joseph J. were hospitalized. Joseph died in a hospital near Dalton, and Henry and James were transferred to an Atlanta hospital, where they also died. Four of the five brothers died of disease before they saw any combat. The fifth, John, served with the company until captured December 16th, 1864 near Franklin, TN. He was sent to the military prison at Louisville, KY, and transferred to Camp Chase, OH on January 4th, 1865. He died there of pneumonia, probably on January 9th, though the prison record states "December 7" which pre-dated his capture.

One can only imagine the sorrow which the Benefield family endured after the deaths of their sons. Their father presented claims for the effects of three of them on May 24th, 1864. Henry's attorney presented a claim for his effects that same day: $41.74 in cash were on Henry's person at his death. The family must have held out hope that John would return to them after the war's end, but their grief was compounded after the next year's holiday season, when they learned of the death of their fifth son in a wind-swept Yankee prison.

Perhaps the most accurate statistic for Ferguson's Artillery is that for prisoners of war, since Federal records are more complete than those kept by the Confederacy. Those records show that sixty-two soldiers belonging to Ferguson's and Beauregard's company were captured and processed as POWs, two men more than once. Twenty-seven of these were captured in April, 1865 at Salisbury, NC and spent only two months in prison before being freed at war's end. Most of the others, captured in 1863 or 1864, either were imprisoned until war's end, or died in prison (five). Several obtained early release with an oath to stay north until the conflict ended, a few men have no disposition noted in the records, and one was exchanged. There are fifteen named men in the battery who opted to join the US forces in return for release from prison, and a few more whose early release suggests they took that option, but the disposition wasn't noted.

Prison was no refuge from hunger, sickness, exposure, violence and premature death, and soldiers sometimes availed themselves of opportunities provided by their captors to exchange the horrors of prison for service in the Federal forces. Eligible POWs were offered, at different times, options to "take the oath" and agree to remain north of the Ohio River for the duration of the war; exchange for a comparable Federal soldier in a Rebel prison, until that system broke down; parole, an early war gentleman's agreement to lay down arms in exchange for release; and enlistment in a Federal unit.

The latter was offered in various permutations. Confederate prisoners were first offered enlistment as sailors, an option chosen by eight of Ferguson's men, most imprisoned at Rock Island, IL. James HC Decker served in Ferguson's almost two years until his capture after Missionary Ridge, and opted to join the US Navy after two months at Rock Island. His application for a US pension made no mention of his prior Confederate service, but he claimed deafness (likely from serving the guns of Ferguson's Artillery), and a scalp wound sustained during his participation in the Federal assault on the southern bastion of Fort Fisher, NC.

The Federals later offered to prisoners enlistment in units of the US Volunteers, infantry mostly designated for western frontier duty against the Indians, thus insuring that former Confederates would not be fighting against and in proximity to their recent southern comrades-in-arms. Four of Ferguson's men joined the 2nd or 5th US Volunteers from Rock Island and Camp Douglas, IL in the spring of 1865; two were rejected for US service, and two others joined cavalry and artillery units. Wellington XL Keefe of Marion Co., SC, a twenty-two year-old fireman, was captured at Ringgold, GA, and spent eleven months in Rock Island before joining Co. F., 2nd US Volunteers. The $100 bounty he earned at enlistment was paid by the Pennsylvania town which was credited for having provided him as a recruit. This ploy allowed states to offset their required quota of enlistees by claiming a Confederate prisoner with the bounty, and served to decrease opposition to the US draft laws, which was particularly strong in PA.[2] When not in the hospital, Keefe spent his service in the west cutting sod and performing the other mundane chores of a US private. It was preferable to the horrors he suffered in the Rock Island camp, and Keefe mustered out of US service on Nov. 7th, 1865 at Fort Leavenworth, KS.

These southern enlistees into the US forces came to be derisively known as "galvanized Yankees." Just as galvanization changed the color of sheet iron with a thinly deposited layer of a non-rusting metal, a galvanized Yankee went from a Rebel gray to a Federal blue uniform. His Unionist peers regarded the conversion as superficial, and his Secessionist peers considered it self-serving.[3] In any case, the newly minted Yankee often served faithfully, as in the cases of Decker and Keefe of Ferguson's Artillery.

Part II. Battle Chronology

An accurate list of the battles in which Ferguson's fought is elusive due to the vanishingly rare "after action reports" for artillery units in the Army of Tennessee (AOT). None exist for Ferguson's in the Official Records (OR). No diaries, and few postwar accounts from the unit's soldiers have come to light. It is easier to list the places visited by the battery, since the OR indicates with which military organizations it was associated. The battle chronologies available in other published works are in fact such lists of place names, but one cannot be certain that the unit was present and engaged in each of those battles. An example of this is the southern disaster at Franklin, TN, where Ferguson's was present with the Army, but went unused along with most of the artillery. The best that can be said is that Ferguson's was present with the AOT at the time of a particular battle, and that its combat role is unknown unless specific information is gleaned from the historical record. Therefore, the chronology which follows understates the unit's battle experience. There must have been many other instances in which its guns were fired, but without contemporary reports or published recollections, they are lost to history.

8. The Campaign at Jackson, MS

In the spring of 1863, Confederate Lt. Gen. John Pemberton's Army of Mississippi was being pushed by Federal forces toward a defense of Vicksburg, and he appealed for reinforcements. On May 2nd Sec. of War Seddon directed Gen. G.T. Beauregard to send 8,000 or 10,000 men from his department to Pemberton's relief.[1] Beauregard replied the next day that he would send 5,000 infantry and two batteries.[2] As part of this force, his Special Orders No. 98 that day directed Brig. Gen. States Rights Gist to take command of four regiments or battalions of infantry and a field battery of four guns from the First Military District of SC, and proceed without delay to Pemberton at Jackson, Mississippi.[3] Thus, Ferguson's joined the 16th and 24th SC, and the 8th and 46th GA to constitute Gist's Brigade.[4] Beauregard telegraphed Pemberton on May 5th that he had sent two brigades of his best troops, Gist's and W.H.T. Walker's, "also two excellent batteries, in one of which is my son, Lieutenant René."[5] The battery likely left Charleston in two contingents: on May 22nd, a communication from Brig. Gen. Ripley to Brig. Gen. Jordan suggests that one detachment of Ferguson's had already left for field service, and the other was preparing to follow.[6] The company took possession of its four Napoleons on May 12th at Montgomery, Al, on its way to Mississippi. These guns were likely of Confederate manufacture, not captures, since they were subsequently described by the Union as of southern origin.[7]

The Campaign at Jackson, MS

It took some time for Gist's brigade to reach Jackson. Its forward elements arrived to fight in the action of May 14th at Wright's farm, but Ferguson's had not arrived in time for the battle, as claimed by other researchers.[8] Gist told Gen. Johnston that he arrived by train in Brandon, Mississippi on the afternoon of May 14th, and when leaving there that day for Jackson, he met another train arriving with Captain Ferguson's battery.[9] He confirmed to Gen. Beauregard that the company was not engaged during the Federal assault and capture of Jackson on the 14th: "None of the troops from your department reached Jackson in time for the affair at Raymond, and only two regiments of Gen. W.H.T. Walker's brigade, Martin's battery, Twenty-fourth South Carolina, five companies of the Forty-sixth Georgia and Eighth Georgia battalion arrived in time to participate in the skirmish and evacuation of the city...The only troops of my brigade engaged at Jackson were those mentioned above..."[10]

Gist's reconstituted brigade encamped forty-four miles from Jackson at Forest Station. When the victorious Federals abandoned the city after destroying its military assets, Gist reoccupied it. By May 25th, Gist was four miles from Canton, north of Jackson, in command of Johnston's 12,000 man army, and Johnston was in Jackson proper. The Confederate forces were planning to consolidate and go to Pemberton's relief at Vicksburg, and the men were constantly marching and countermarching. Communicating with Gen. Beauregard, Gist said "Ferguson's battery is in fine condition, and your son has just left my quarters in excellent health."[11] The unit was then encamped at Vernon, MS, just south of the Big Black River, about ten miles from Yazoo City. Gist's brigade was assigned to Gen. W.H.T. Walker's division. Lt. Col. Capers of the 24th SC characterized the months of May and June as a test of his brigade's "spirit and discipline" in the summer heat, "badly supplied with shoes and actually, at times, suffering for water fit to drink."[12] In Ferguson's company, exposure to the weather was something to be endured, as noted by a letter from William Chastain to his sister on June 12th, 1863 from Yazoo City, MS.[13]

> Dear Sister it is with pleasure that I take my pen in hand to inform you that I am well at present hoping when these few lines comes to your hand they may find you enjoying the same blessing I received a letter from you the other day and was glad to hear you are well I hav no good news to rite you at present I heard last night that we had a hard fight at Vicksburg and we whip the Yankees bad and hold a great many of them but I don't know how true it is We expect to hav a hard fite here in a few days We have hard times out here no tents we hav to lay out in the rain and take all sorts of bad weather and you know that is bad but we make out toleable well we cant have everything that we want in the war I want you to tell all the pretty girls howdy and tell them I hope I will get to come home and see them So I must close by saying I remain your loving brother until death
>
> Direct your letter Yazoo City --- Walker Division Gist Brigade in care of Capt Fergerson L Battery

Johnston's army was thwarted in his plan to rescue Pemberton at Vicksburg due to the Federal's superior numbers and positions. When Vicksburg fell on July 4th, the Confederate Army of Relief fell back to the fortified lines around Jackson to await the expected attack of Grant's forces, now freed from their siege duties at Vicksburg.

The US forces, under Sherman, arrived at the outskirts of Jackson on July 10th to 11th. Walker's division, including Gist's brigade with Ferguson's Artillery, occupied a position on the left center of the Confederate line, the brigades of Ector, Gregg, Gist and Wilson in order, with the division's right on the Clinton road (Fig. 1). They were at the extreme western extent of the Confederate's convex line of fortifications around Jackson, near Ft. Johnston, which mounted a 32 pounder gun. In their front, astride the Jackson road, was the XV Corps of Sherman's army with Preston Blair's 2nd Division on the left of the road, and Steele's Divisions on its right.[14]

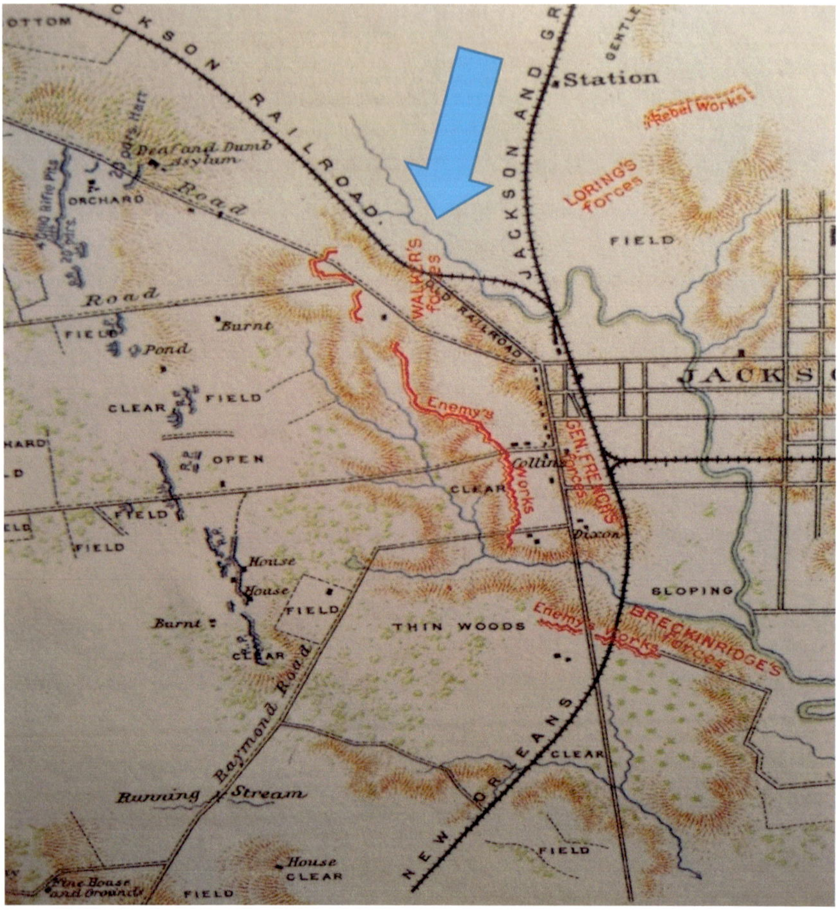

Fig. 1 Walker's forces (Ferguson's Artillery) at Jackson, MS. *The Official Military Atlas of the Civil War*, Plate 39, 1.

US batteries were placed on the Federal front, including four 20 pounder Parrott rifles on the grounds of the Deaf and Dumb Asylum, only 400 yards from the Confederate line[15] (Fig. 2). Entrenched southern units faced a similar situation around the entire ring of Jackson's fortified line: massed Union troops and artillery testing the strength of the defenses with piecemeal infantry probes, prolonged cannonading, and constant sharpshooting (Fig. 3).

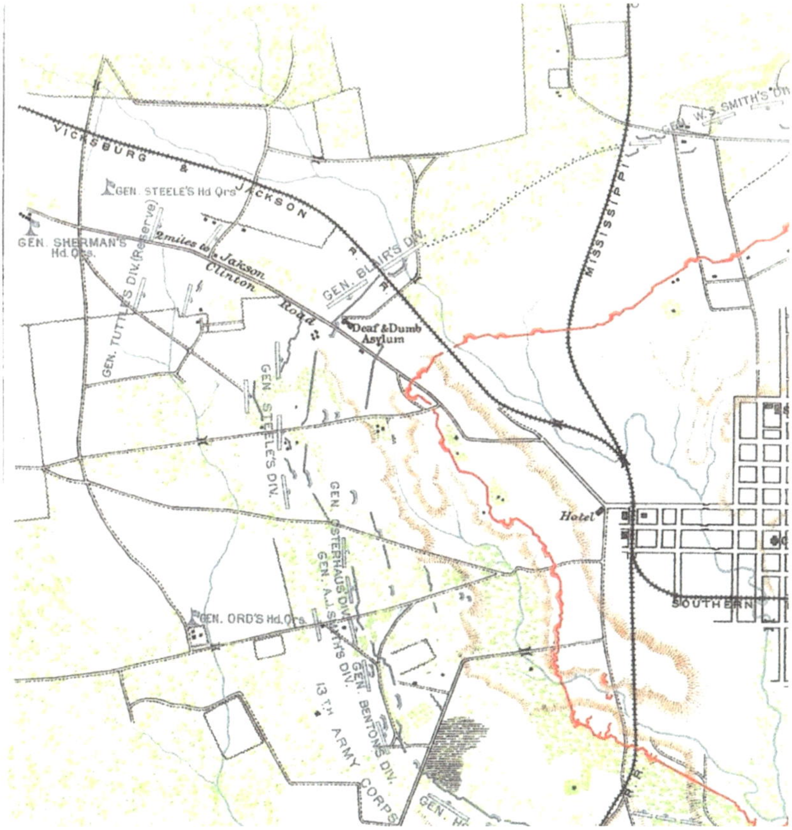

Fig. 2 Blair's and Steele's (US) Divisions astride the Jackson (Clinton) Road.
Official Military Atlas of the Civil War, Plate 37, 5.

Fig. 3 Confederate lines around Jackson, MS.
Official Military Atlas of the Civil War, Plate 37, 2.

While the artillery barrage was impressive, with up to three thousand rounds fired into the city in one hour on July 12th, the sharpshooting apparently exacted a greater mental toll on the besieged southerners.[16] Flavel Barber of the 3rd TN wrote July 14th: "sharpshooters [are] still annoying us constantly. A person cannot step out of the trenches without danger. Men are shot every day and every hour."[17] Lt. J.M. Armstrong of the TN Lookout Battery wrote to his sister that same day of "constant skirmishing on both sides all day & night. One man was shot through the breast. He did not die but was expected to."[18] This might well have been a reference to Captain T.B. Ferguson, who was commanding the artillery of Walker's Division. On July 14th, while executing an order from Gen.

Walker, he was shot with a Minié ball through the upper lobe of his right lung, putting him out of action and prompting news of his death back home in Charleston. The report proved to be erroneous, but the wound would require a long convalescence. In Charleston the press responded: "We regret to hear that Capt. Thomas B. Ferguson, the well-known commander of the light battery recently here, has been seriously though not fatally wounded during one of the late battles in the West."[19]

Gen. Johnston was forced to abandon Jackson when the arrival of a Federal train with more ammunition for the heavy siege guns, and the appearance of more Union gun emplacements, suggested that Sherman was preparing for a protracted bombardment of the city.[20] Johnston made plans for a night-time evacuation, which he termed a "retrograde movement." He ordered that beginning at 9 pm on July 16th, "the artillery in or near the trenches will be moved by hand for at least half a mile to the rear," this laborious task aided by details of twelve infantrymen for each piece, "who, with the cannoneers, will drag them to the points where they are to be limbered up."[21] Walker's Division and Ferguson's moved silently through the city and crossed safely over the Pearl River using a pontoon bridge at Carson's Ferry, the remainder of the army leaving in a similar fashion. By sunrise on July 17th Johnston was gone, and the surprised Federals were able to enter Jackson and continue its destruction. Johnston moved to Morton, pursued as far as Brandon by a fraction of Sherman's forces, which broke off the pursuit after burning a quarter of the city and terrorizing its citizens.[22]

From the time it left Charleston in May to the end of the Jackson campaign, the company lost five men to capture, four to death from disease, and their captain to a severe wound. Henceforth, the unit would be led by René Beauregard.

9. Tennessee in '63

Ferguson's Artillery followed Johnston's army into camp around Morton, MS. On August 20th, 1863, the unit was still attached to Gist's Brigade in Walker's Division, Department of MS and East Louisiana.[1] Lt. Beauregard took the opportunity to refit his battery, drawing clothing, additional horses, mules and harnesses, discovering to his distress that much of the issued equipment and many of the animals were unserviceable and merited only an immediate return to the quartermaster. After five weeks at Morton, Walker's Division was ordered Aug. 23rd to reinforce Gen. Braxton Bragg at Chattanooga, TN for temporary service in what would lead to the battle of Chickamauga.[2]

Due to the sorry condition of the southern railways, the trip was not a quick one. Johnston was loathe to part with any troops, and informed Bragg Aug. 24th that he "sent 9,000 infantry and artillery. This road works so wretchedly that it will take five days, including yesterday to get them off. This is a loan to be promptly returned."[3] Lt. Beauregard's unit headed east toward Atlanta, via Montgomery, AL, where he drew another 23 horses and telegraphed his father Sept. 3rd that he was "On way to Chattanooga with battery" (Fig. 1).

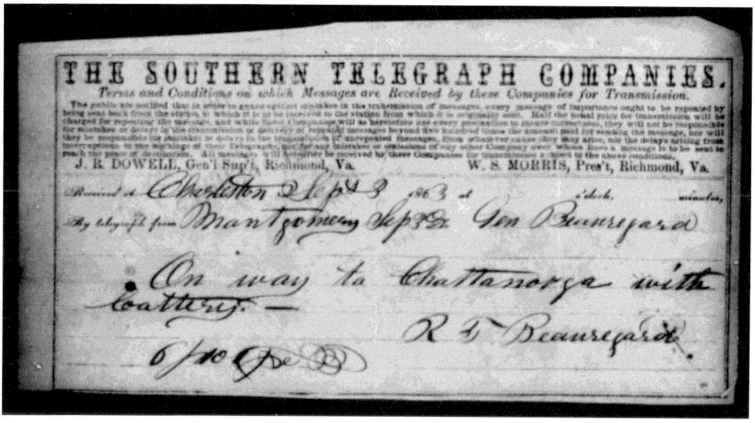
THE SOUTHERN TELEGRAPH COMPANIES.

Terms and Conditions on which Messages are Received by these Companies for Transmission.

J. R. DOWELL, Gen'l Sup't, Richmond, Va. W. S. MORRIS, Pres't, Richmond, Va.

Received at Charleston Sept 8 1863 at o'clock, minutes,

By telegraph from Montgomery Sept 8 To Gen Beauregard

On way to Chattanooga with battery.—

R. T. Beauregard

Fig. 1 Telegram from Lt. René Beauregard to Gen. Beauregard.

On Sept. 4th Walker was ordered to Rome, GA to be ready to move by rail in any direction. His artillery was to be stopped at Atlanta, and redirected to join the infantry at Rome.[4] On that day, Beauregard was in Atlanta, drawing forage for the animals, and headed toward Rome per orders. Walker's Division, part of the Reserve of the AOT[5] moved out from Rome to join the right wing of the Confederate army and fought at Chickamauga Sept. 18th-20th, but Ferguson's Artillery remained behind at Rome with the 16th SC INF and a cavalry force, guarding the army's left wing. There were insufficient train cars to transport Ferguson's and the 16th SC from Rome, and the battery did not rejoin Gist "until the morning of Sept. 23, three days after the battle."[6]

On Oct. 7th Beauregard requisitioned various items of clothing for 106 men, and three days later drew forage at Chickamauga for 106 animals. This suggests that his company was well-manned and equipped for the coming campaign around Chattanooga, where the unit was to be part of the besieging forces. The officers had been so successful in recruiting that there were more men than could be accommodated in a single company, and Henry Calhoun was working to get a second company equipped. Private James A. James was one of those soldiers, and wrote his father from Atlanta on Oct. 20th, 1863.

Tennessee in '63

My Dear Father,

I take this opportunity of writing to you to inform you of my whereabouts & welfare. My health is very good at present. We are stationed at Atlanta, Ga. for a few days. I do not think we will stay long. When we leave here, we will go to Chattanooga, Tennessee. There is nothing new about this place worthy of communication. Atlanta is quite a lively place now. There is a great many sick & wounded soldiers here. There is eight hospitals in this place & they are all full of sick & wounded. They all get very good attention, I believe.

There has been a great many Yankee prisoners here but they have sent a good many off & a great many have died & I would not care if all the rest would follow suit. There is not but few here now except those that are not able to be sent off & I don't care if they never get able. You may think I am hard - maybe, but it is so. They deserve death and let them die.

Brother John & John Chapman left camp on the 13th of this month & have not heard from them since. I expect they have gone home. They left in the night about 10 o'clock. They hunted for them next day but did not hear from them. I want to go home as bad as anybody but I will never go if I have to desert to go because I would rather die than to be called a deserter.

Times is hard here. We do not get much to eat. We have not got any bread for dinner today nor will not have any for supper, nor breakfast. We do not get much to & have not got anything to cook what we do get in.

We have not drawn any money yet nor I don't know when we will. You must give my love to Susan, Jane & family & all the rest of my sisters & friends. I want you to write to me immediately & direct your letters to me in Capt. Calhoun's Light Artillery, Ferguson's Battery,

Atlanta, Ga. Nothing more but remain your affectionate son, - J.A. James[7]

James and the rest of Ferguson's did leave for Chattanooga soon after his letter, and by Nov. 9th were posted with Walker's Division on the left of the Confederate line, from the eastern base of Lookout Mountain to Chattanooga Creek[8] (Fig. 2).

The Federal forces decided to break out of Chattanooga in November, and on the 23rd overran Orchard Knob, in front of the western slope of Missionary Ridge. In response, Gen. Bragg recalled his outlying forces and that night withdrew Walker's Div. from its Lookout Mountain post, repositioning it to the right of his line on Missionary Ridge, the first of the Confederate units to occupy the heights.[9] This further weakened the tenuous defending forces of Lookout Mountain, which on Nov. 24th were easily swept aside by the advancing Federals, in what became known as "The Battle Above the Clouds." That night the remaining Confederates below Lookout Mountain withdrew to Missionary Ridge, there to await the expected onslaught of the entire Union army.

Col. James Cooper Nisbet, of the 66th GA INF, in Walker's Division, recalled moving during the night from Lookout Mtn. to the north end of Missionary Ridge. There, Ferguson's battery of twelve pounder Napoleon guns had to be dragged up to the summit, which was accomplished with the use of ropes and pulleys manned by the artillery crews and infantry. "Lt. Beauregard, commanding the Artillery Company, reported that his horses were too weak to pull the guns up. I detailed a company of infantry with each piece; some of the men pushed, others pulled, using ropes, and at last we were all on top. When we reached the crest, it was still dark. We moved to the right, slowly, over a dim, wood-road. At several points we had to make a road for the artillery."[10] The battery occupied a position in the right of the Confederate defenses, three quarters of a mile to the south of the entrance of the Nashville and Chattanooga RR tunnel through Missionary Ridge at Tunnel Hill[11] (Fig. 3, 4). Nisbet and Gen. Gist, commanding Walker's Division, helped Beauregard post his guns

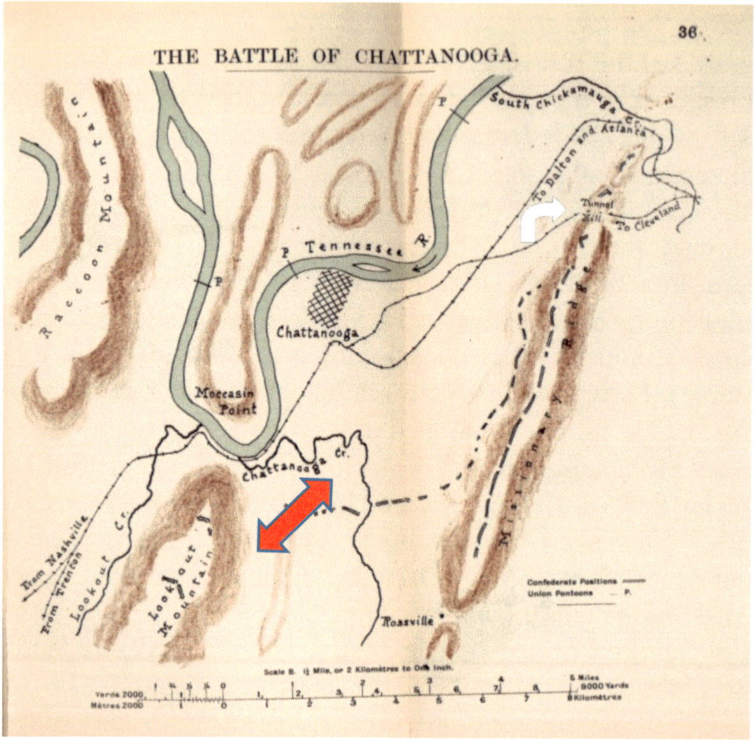

Fig. 2 Tunnel Hill (white arrow) and Walker's Division (red) to November 23rd.

on a commanding point so as to be able to fire over the heads of Nisbet's men. At an elevation of 1080 feet, attacking Federals would have had to climb over 300 feet to reach Beauregard's position.[12] Thus, Ferguson's and Walker's Division were not directly assaulted during the battle (Fig. 5, 6).

Gen. Sherman's Army of the Tennessee was tasked with attacking Missionary Ridge from its northern extent, to hopefully roll up the Confederates along their right flank, sparing what was felt to be an

Fig. 3 Missionary Ridge at Tunnel Hill, looking east.
Author photo.

impossible frontal assault along the length of the defenses. At daylight on November 25th, Nisbet observed Sherman's advance and ordered Beauregard to open fire. The lieutenant asked "Have you an order to that effect?" and Col. Nisbet replied "Commence firing Lieutenant; what we see is order enough for me." Beauregard continued long-range fire for the next six hours, targeting Federal forces approaching his front and from the direction of the Glass farm toward Tunnel Hill. This right section of the Confederate line did not yield to the Federal assault that day, thanks to spirited defense and the formidable nature of the terrain. Ferguson's Artillery lost Private William Crookes, KIA. Unaware that the Confederate center had collapsed and believing that their fighting was over for the day, the battery was cooking supper when Gen. Cleburne rode up to Nisbet and Lt. Beauregard, and told

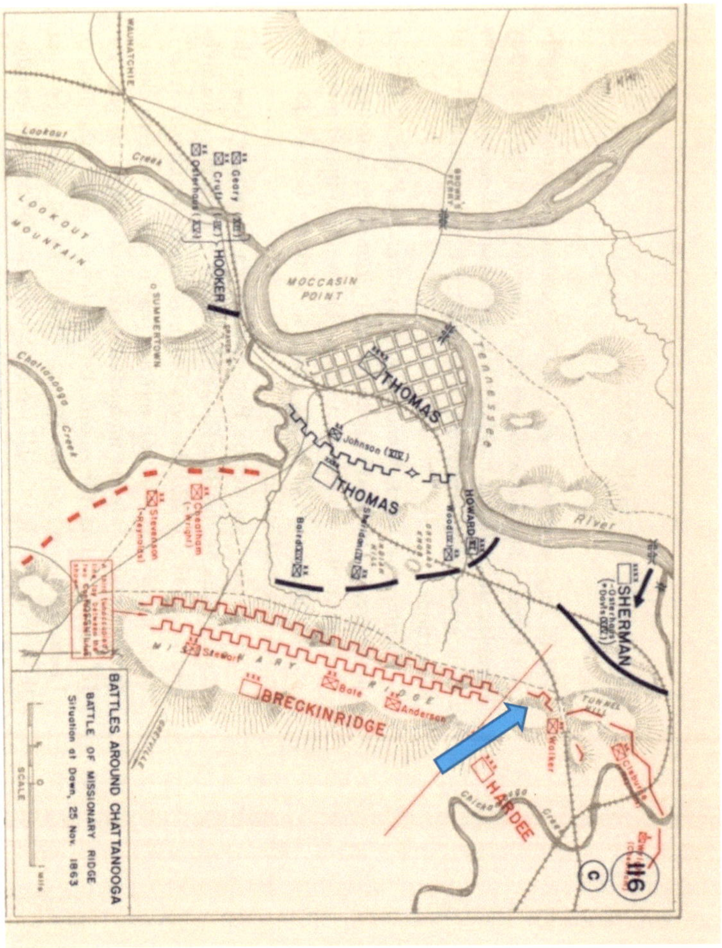

Fig. 4 Walker's Division (Ferguson's) south of Tunnel Hill, Missionary Ridge. Arrow shows Ferguson's position. West Point Atlas, Map 116, Esposito, Vincent J., United States Military Academy.

them that the rest of the Army was retreating, and that they should retire down the eastern slope of Missionary Ridge to Chickamauga Station. Recalled Nisbet:

Young Beauregard was much worried. Said he did not know of any road to get off the ridge; that his old, poor horses could hardly stand up on level ground, and that he was going to lose his fine guns. In fact he was so much troubled the tears came in his eyes. I said to him that I will stand by you; I have some men hunting for torch pine. We will have torches soon. I will keep my own regiment with your battery. Torch-bearers arrived and found a 'sorter-of-a-little, old, blind wood-road' down the eastern side of the ridge. I detailed a company to go with each gun; with orders to lock all the wheels with ropes or chains, and to hold the carriages back by hand off the feeble horses...The boy captain Beauregard got in better spirits when we arrived safely at the foot of the ridge. There, crossing a muddy branch, the horses on the first gun lay down in the creek and all the profanity and beating of their driver could not move them. A company of infantry was detailed to push guns and horses out of the icy water. 'It is wonderful what eighty-five or ninety men can accomplish when they put their shoulders to the wheel.'

The company proceeded at a gallop down the slope to the temporary shelter of Chickamauga Station on the Western and Atlantic Railroad. The exhausted, famished soldiers of Nisbet's brigade and Ferguson's joined in a free-wheeling "grab game" of the available rations, filling their hats from a barrel of sorghum syrup and wringing it out of the clothes of those men unfortunate enough to have been tripped into the barrel by a mischievous comrade.[13] As the remaining elements of the defeated Confederate forces straggled to the rear, the artillerymen would have gotten what sleep they could.

Bragg moved his army at daylight on the 26th, marching toward Ringgold, GA. The march from Chickamauga Station was slow, as the weather had reduced the roads to mud pits, miring the wagons and artillery carriages and exposing them to capture by the pursuing Federal army. The more mobile infantry forged ahead of the wheeled forces, which were further slowed by branches of the sinuous Chickamauga Creek. Gen. Gist, commanding Walker's division, covered the rear of Breckenridge's corps[14] and by nightfall had travelled over the South Chickamauga Creek at Graysville, where they heaved one of Ferguson's probably damaged Napoleons into a ravine (Fig. 7).

"The following circumstances occurring at this point were related by Orderly Sergeant W. H. Bruton, of Company A, 6th TN INF. The brigade found the bridge across the Chickamauga River at Graysville burned, and that deep little stream swollen. The night was dark and bitter cold. Close in their rear could be heard the dull rolling of artillery carriages and upon either side the enemy's cavalry could be heard taking position on the rocky roads which ran parallel. Gloomier still, the weary soldiers could hear the splash and plunge of their own artillery as it was abandoned and rolled into the river. More gloomy than all of this, they could hear men riding off on the artillery horses, evidently bent on escaping a pressing peril."[15]

The vanguard of the Federal pursuit, moving southwest along the Federal Road (the modern Hwy 41) crossed Peavine Creek in the dark and surprised Ferguson's along the Graysville road (Fig. 8).

"Col. William Stoughton advanced undetected eastward toward Rebel campfires along the Graysville and Ringgold Road. Attacking from ambush, they fired a volley into T.B. Ferguson's SC battery and its infantry supports of the 41st TN and 16th SC. Howling like wolves, the Federals charged capturing three cannon of the battery, limbers and caisson, their flag, and about 110 prisoners including killed and wounded. The cannon were mired in the mud."[16]

Fig. 5 Plaque on Missionary Ridge near the intersection of N. Crest Rd and Marshall St.

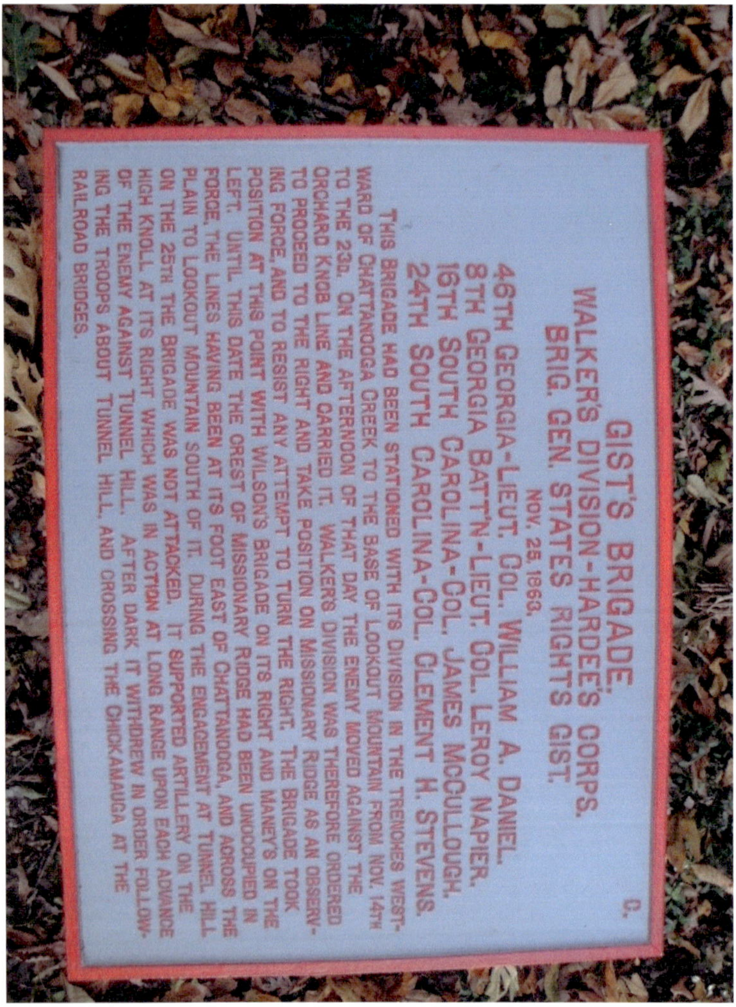

Fig. 6 Plaque on Missionary Ridge near the intersection of N. Crest Rd. and Marshall St.

Fig. 7 South Chickamauga Creek at Graysville Rd bridge. Author photo.

It was likely this volley which severely wounded Private Archibald Young, fracturing his thigh and leading to his death in a Federal hospital the following March.

Col. Stoughton, 11th MI, commanding a Federal brigade, reported the action as follows.

"Soon after dark we crossed a small creek 1 mile south of Graysville...and moved through a piece of woods to attack the enemy on a road leading from that place to Ringgold. On approaching the road we surprised and captured his pickets, and learning his position moved promptly forward and made a vigorous attack, capturing over 60 prisoners, 1 flag, 3 pieces of artillery, 2 caissons with horses and equipments complete. One gun dismounted and thrown over the riverbank and two caissons abandoned by the enemy in his flight were found the next morning and secured. The captured artillery was known as Ferguson's battery."[17]

Private George Pretts of the 69th OH recounted the affair:

"We moved along in that manner for some time and in single file, for we had gotten off of the road into the woods and each man must follow his comrade in front. Finally the foremost regiments halted and bunched up, and as we in single file came up, we took our places...The order was then given to front, and when we did so and moved a few yards there was quite a bank looming over us. It must have been fill for a railroad. It was about 25 feet at the base and about 14 feet high. We could see the reflection from fire on the opposite side. The order was given to charge and not fire unless ordered to do so but when we gained the top of the embankment the temptation was too great for nearly one-half the command and they cut loose on the enemy. It seems that in our immediate front there were three or four small fires, which were surrounded by several chill men in the gray garb of artillerymen. There was but one slight volley fired. Then everything was as still and calm as around some country farmhouse where war had never been. Immediately after the firing our Colonel shouted, 'Cease that firing.' The result of our sneak was the capture of the famous South Carolina battery and 110 prisoners. The Sixty-ninth Ohio formed a square and marched the prisoners back some distance and went into camp for the balance of the night... After the volley was fired the Rebs came running up to us in surrender...The ones who seemed to want to make their escape would jump from tree to tree. We could see them move and command them not in the sweetest terms ever uttered to halt and come forth and throw up their hands and say 'I am your prisoner' which in most cases they did."[18]

Fig. 8 Pea Vine Creek and the road to Graysville (Greysville). Arrow shows Graysville road on which Ferguson's lost three of its guns. US Army Topographic Engineering Center. [Map of Catoosa County and parts of Whitfield, Murray and Walker counties, Georgia] Chattanooga, Sept. 12 [1863?]. https://www.loc.gov/item/2008628277/

The news of the engagement was also reported in the SC press:

"Unfortunately, Ferguson's battery of four guns, belonging to Walker's division, was captured. The horses were in very bad condition, and unable to keep up with the column; hence the disaster. The greater part of the men and horses escaped. You will be astonished to hear that the horses in the artillery service, the most important in the whole army, are the most neglected - a fact, however, which quartermasters, and even artillerists, seem incapable of comprehending. The loss of the battery occurred soon after dark and not far from Ringgold...Those companies of the 16th South Carolina, acting as a special guard to Ferguson's battery, were dispersed and many of the men taken prisoner."[19]

Ferguson's Artillery Company had lost its four bronze Napoleons for the first time (Col. Nisbet had incorrectly recalled that they had six at Missionary Ridge, and that the guns had been captured from the Federals earlier at Chickamauga). The guns were lost through no fault of its own. They had no chance to defend themselves from the sudden Federal surprise attack, and their infantry support offered no resistance. Seventeen men on the company's muster rolls are listed as captured at Graysville or Ringgold GA on Nov.26th and 27th, 1863. The men who escaped into the woods straggled back and joined Gist's Brigade on the road to Ringgold, toward its eventual winter quarters in Dalton. Beauregard took a short leave to visit his father in Charleston over Christmas, and rejoined the battery as it licked its wounds and regrouped.[20]

Dr. Bull took the opportunity to write to the families of the wounded soldiers, as in the following letter to Archibald Young's wife.

Charleston Jan. 16, 1864

Mrs Margaret Young

Dear madam,

Your husband was wounded at the battle of Mission Ridge on the 26th of November. He is shot in the leg halfway between the hip and the knee, and the bone broken. He is a prisoner in the hands of the enemy, and will remain so until exchanged. When I left him on the 29th of December he was recovering rapidly and in all probability will be well in three or four weeks. You will find enclosed fifty dollars which he requested me to send you.

Yours respectfully

Wm Izard Bull

Surgeon Ferguson's Battery

Young wrote his wife soon after, as follows.

Gen'l Field Hospital
Chattanooga, TN Feb 13th/64

Dear Wife,

Being as there is an order just come in for the removal of our wounded across the lines but I am sorry to say that I am not able to accompany them as wound will not allow of my being moved at present so I thought I would avail myself of the opportunity of sending you a few lines to let you know how I am getting along I am improving a little but very slow I am under good treatment and I hope I shall be able to come home. When another lot is sent through which I think will be soon I have sent you two letters since I have been here and I feel very anxious to hear from you I sent you one letter by Doctor Bull the doctor of my company I sent you fifty dollars by him also and I should like to hear if you received it or not. And I want if you can get a chance to come see me. And I want you to understand how my account stand in the company. I have seven months extra pay coming to me that I have signed for and I have six months wages coming to me up to the 25th of November if you see fit to come and see me there no obstackle in the way you will get good accomodations and good treatment while you remain with me and you can draw my money if you can get to Atlanta after seeing me.

So I remain

Your husband

A M Young

Young was ministered to by representatives of the Sanitary and Christian Commission, but despite their attentions and the care of his Federal surgeons, his condition worsened and he passed away on March 2nd.[21]

10. The Atlanta Campaign

After the retreat from Missionary Ridge the AOT went into winter quarters at Dalton, GA. Under the new command of Joe Johnston, the Army reequipped and bolstered its morale. Ferguson's was resupplied as part of Martin's Battalion, which also included Bledsoe's (MO) and Howell's (GA) batteries. Chief of Artillery of the Army of Northern Virginia, Brig. Gen. W.N. Pendleton, visited the AOT artillery for ten days in March, examining each battery in detail and witnessing their drill. The six pounders he felt were "nearly useless, if not indeed worse, employing and exposing as they do a number of men and animals, while they can scarcely ever accomplish anything against the more powerful guns or even the long-range muskets of the enemy."[1] On March 29th, 1864 Ferguson's had two Napoleons and two six pounders, replacements for its guns lost after Missionary Ridge.[2] The obsolete six pounders were two of only fifteen in the entire AOT. Chief of Ordnance Lt. Col. Hypolite Oladowski had been working since early 1863 to increase the number of Napoleons in service, and by autumn of that year had successfully lobbied Richmond to direct the Confederate arsenals to give the AOT preference in the recasting of its unwanted bronze six pounders and many of its 12 pounder howitzers into Napoleons.[3] Losses in Ferguson's had been twenty-one men captured during the retreat from Missionary Ridge (compare to the seventeen documented on its muster rolls) and fifty-six horses. The battery was down to ninety-five men and five officers present for duty, and needed an additional twenty horses, but except for the lack of some few equipments was considered in good condition.[4] Gen Johnston reported that the artillery horses of his army "were so feeble that in the event of a battle we could not hope to maneuver our batteries, nor in case of reverse to save our guns."[5] Mules were increasingly used in the artillery teams to supplement the deficient horses, despite the resulting inefficiency.[6] Ferguson's added ten mules to its inventory by the end of February. The supply of long forage (hay) was scarce, and the horses had been suffering since Missionary Ridge.[7] Gen. Pendleton estimated that by April 10th Ferguson's six pounders would be replaced by two more Napoleons, and that five hundred more horses would supplement the AOT artillery at an early date. By April 12th, the company had its 6 pounders replaced, as Oladowski had just received Napoleons from the Macon and Augusta arsenals, and Ferguson's was back to its original complement of four, light twelve pounder guns.[8]

This was the first winter Ferguson's spent in field quarters. By this time, the artillery of the AOT had been reorganized. One battalion of three batteries was now attached to each division in a corps, and each corps had an artillery command; there was also a reserve artillery with its own command.[9] During the winter of 1863, Ferguson's was attached to Walker's division in Hardee's Corps. Descriptions of the winter quarters at Dalton suggest that the battery was probably comfortable. Gist's brigade was quartered two miles east of Dalton on the Spring Place road, the camp described as a "beautiful one, very nice, snug houses - wide streets regularly laid off and looking like a village – the men well clothed and shod – in excellent discipline and well drilled."[10] Ferguson's Company was at Spring Place, on the left hand side of the Spring Place road leaving Dalton, seven miles from the city (Fig. 1).

Private John A. James wrote to his parents January 18th, 1864, with travel instructions to visit him via the rails from Greenville, SC to Columbia, south to Branchville, then west to Augusta, GA and

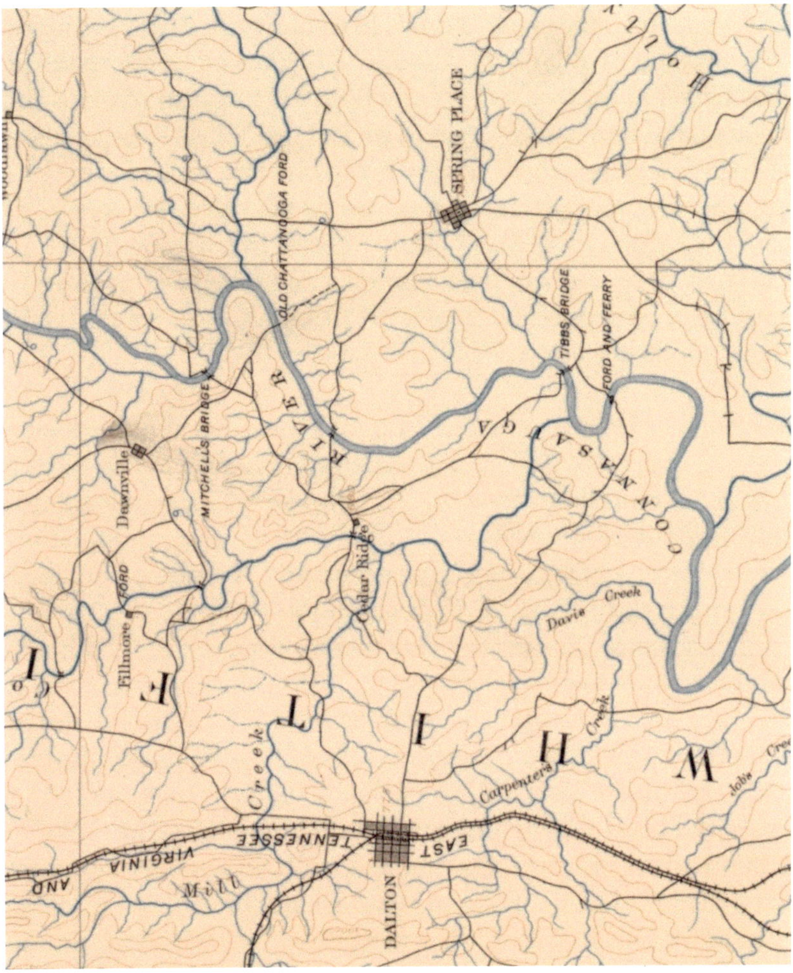

Fig 1 Ferguson's winter quarters at Spring Place. Geological Survey, U.S., Henry Gannett, Gilbert Thompson, Samuel S. Gannett and F.M. Pearson. *Georgia, Dalton sheet* (Washington, D.C.: U.S.G.S., 1886), https://www.loc.gov/item/2007626717/.

Atlanta, and north to Dalton. He reported that:

"times is hard here and worse a coming...It is very muddy too...I want one pair of drawers, one pair of pants, two pair of socks, one pair of gloves, and a hat and a coat. You need not bring no scarf. Fetch some fifteen pounds of flour and some dried fruit and some butter and some chickens and I will pay for the flour for I have to pay one dollar for one pound here and am glad to get it at that for we only get enough for two little snacks a day and we eat it at one time. And I want you to bring some red pepper for I want you to bake me one loaf of tight bread. I want one more good bottle of brew from home for we can't cook good here and the beef that we draw they have to lean it up against a tree to shoot it or prop it with rocks. It is one pound of beef and ten pounds of bones."[11]

Sherman's army of over one hundred thousand men probed the AOT defenses west of Dalton as early as February, finding them too strong to take in direct assaults. The imposing heights of the ridges meant that Gen. Johnston could defend them against a numerically superior force, allowing him to concentrate his defenses on the few gaps through the ridges, through which the Federal army would be forced to advance. James wrote his parents again on April 27th, with premonitions of the impending clash. "You say that you have had hard times back in Carolina and I must say there are hard times here as if there are not hard times here I do not know what hard times are. I have little news here but we expect a big fight near soon as the baggage has been sent to the rear and all furloughs stopped and I do not know when they will start again. I really want to get home but I am out of hart of going home...I will close nothing more till death."[12]

In the first week of May, Sherman pressed attacks on the northern ridges and gaps, but was thwarted by the Southerners. One wing of the Federal army was able to pass through the undefended Snake Creek Gap to the south, which threatened to cut Johnston's supply route from his rear, necessitating the abandonment of Dalton and the movement of the AOT south to Resaca. Thus began Sherman's Atlanta Campaign and a summer of fighting for Ferguson's Artillery.

On the evening of May 12th the artillery and infantry marched toward Resaca.[13] It is certain that Ferguson's participated in the battle there, but its location and actions cannot be stated due to its absence in any after-action reports. Private Pat Prince was slightly wounded in the side, the only reported casualty at Resaca.[14] Hardee's corps was in the center of the four mile Confederate line, and presumably Ferguson's was posted there. The artillery chief, Col. Smith, would have taken orders from Hardee and posted his batteries where required, so Ferguson's could have fought with corps units ordered elsewhere on the battlefield. Units of Walker's Division, to which the company was attached, were employed both in the center of the line and to the south at the contested Lay's Ferry crossing of the Oostanaula River. The other two batteries in Martin's Battalion of artillery suffered greatly at Resaca, with multiple casualties.[15] The Confederate army retired to the south in the evening of May 15th, when the left of its line was threatened by Federal units crossing the Oostanaula.

Hardee's corps left Resaca moving south toward Adairsville, GA to join the other two corps of the AOT. On May 16th near Calhoun, GA, where the Oothkalooga Creek crosses the Rome Road, the 2nd Div. of the Federal 16th corps attempted to cut off Hardee's wagon train, and was met by troops of Gen. Walker's division, including the 24th SC (Fig. 2).

They repelled the Federal troops, inflicting about fifty casualties, killing Col. Patrick Burke of the 2nd Brigade.[16] Casualties in the 24th SC numbered forty-one according to Ellison Capers.[17] Hardee's artillery was involved in this action, the specific units not evident in the OR.[18] Ferguson's privates Geary, Griffith and Quinn went missing here, captured between the 16th and the 18th. The circumstances of their capture is unknown, but it may be telling that they all had enlisted at Charleston the same week, that Griffith and Quinn had both previously been court martialed for desertion, and that Geary was to quickly get out of Rock Island prison by enlisting in the US navy. This raises the possibility that the men slipped away during or after the fighting at Calhoun, with the intention of being captured. In any case, Griffith's absence meant the company had lost its only bugler.

Gen. Johnston's strategy was one of repeated entrenching, then retreating southward when the Federals threatened to flank his forces. This resulted in daily skirmishing for both the infantry and artillery arms, the latter deploying to blunt the enemy advance. From Calhoun, to Adairsville, Cassville, Allatoona, New Hope Church, Dallas, (where two men of Ferguson's were wounded), and south along the route of the Western and Atlantic RR, the AOT waged a fighting retreat through the month of May. By June, Ferguson's, with two other batteries and Bate's division, occupied a position atop Pine Mountain, between Kennesaw Mountain and Lost Mountain, just over a mile forward of the main Confederate line. Its isolation made it an inviting target for the Federals, who engaged the Confederate salient for over two weeks. The battery received clothing there on June 7th (Fig. 3).

On June 14th, Gens. Johnston, Hardee and Leonidas Polk were observing the Federals, and attracted fire from a battery not six hundred yards distant, reportedly personally directed by Gen. Sherman. Polk, standing behind Ferguson's emplacement, was nearly cut in half by the third round from a rifled piece (Fig. 4). The battery witnessed the death of "Bishop Polk," whose loss was keenly felt throughout the AOT.[19] The Confederates abandoned their advanced Pine Mountain position that night, and the next day were assaulted by the Federals along the main line, forcing Johnston to entrench on another line anchored on Kennesaw Mountain.

Private James was hospitalized and missed the death of Polk. When he wrote home on June 20th, 1864 his previous eagerness for a visit from his family had been replaced by an indifferent ennui.

Dear Father,

I take this opportunity of writing you a few lines to let you know how I am. I am as well as could be expected. Received your letter

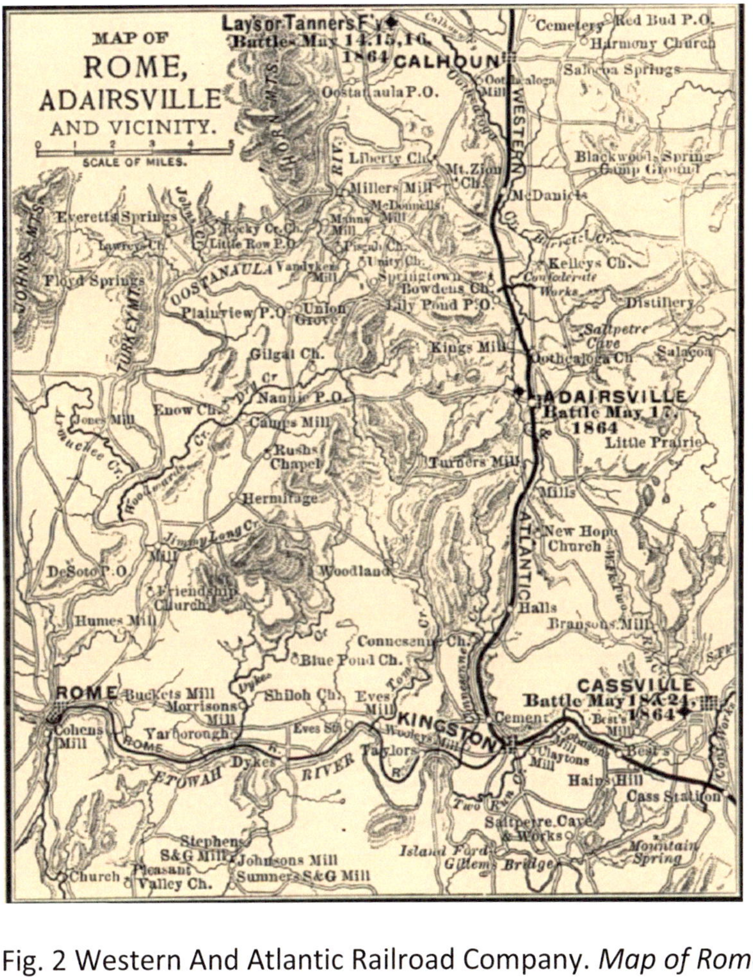

Fig. 2 Western And Atlantic Railroad Company. *Map of Rome, Adairsville and vicinity*. (Buffalo, N.Y.: Matthews, Northrup & Co., Art-Printing Works, 1864). https://www.loc.gov/item/99447298/

Fig 3. Arrow indicates Ferguson's position and Polk's death on Pine Mountain. Blakeslee, G.H. *Map of the environs of Pine Mountain, Lost Mountain, Kenesaw Mountain and Little Kenesaw Mountain.* [1864] Map. https://www.loc.gov/item99446392/

Fig. 4. Alfred R. Waud (1864 June 14) *Death of Gen. Polk, Pine Mountain, Kenesaw.* https://www.loc.gov/pictures/item/2004660330/

this morning and was glad to hear from you and to hear that you was all well.

There is no news to write — only they are all fighting yet up here and bunches of wounded are coming in here every day and men are dying off fast here. There are from 10 to 15 taken out of the dead house every day.

It has been raining for 4 weeks here every day. You wanted to know if you could come through to this place. You can come if you want to. There is nothing to hinder you. There is folks come further than that. If you will try, you can come without any trouble.

When you write again, you will let me know whether fruit is ripe or not and whether there is any of it there or not. It is unripe in this country and if any of you come, I would like for you to put some in some packet and fetch it to me. You can come if you want to and you can do as you wish about it. Come if you want to come and if you don't want to come, don't come. I would be very glad to see you all and stay with you awhile and take supper and perhaps a sit with you. The folks is mighty indifferent here. Nobody cares for me and I care for nobody.

Tell Jane and all the children howdy for me and that I would like to hear from them all. Give my best respects to all my friends...I don't know whether I will ever get home or not but I hope I will get to come home soon. I hope you will write as soon as you get these few lines.

When you write, you must let me know how the times is out there — whether they are any better there than here or not, for it is hard times here and looks worse a coming. You must write soon as you get these few lines. Give my love to all the friends and a share for yourself. When you write, direct yours letter to Fairground Hospital, Ward No. 19, Atlanta, Ga.[20]

His comrades back on the "third Kennesaw line" were on the strongest so far, and Sherman chose a direct assault rather than his previous gambit of flanking Johnston's army out of its entrenchments. At least two Ferguson soldiers were wounded here on June 17th.[21] On June 20th a Federal probe of the Confederate left was repulsed by two divisions of Hood's corps at Kolb's Farm. On the 27th, a Federal artillery barrage was followed by a frontal movement against Cheatham, Cleburne and Cockrell, costing three thousand US and one thousand CS casualties.[22] Ferguson's was posted with Hardee's corps somewhere to the right of Cleburne on Kennesaw Mountain, able to put an enfilading fire into the left flank of the attacking Federals.[23] The battery received another clothing issue on that day. Having failed in a direct assault, Sherman returned to his flanking maneuvers, and on July 2nd Johnston abandoned the Kennesaw line, retreating to Smyrna, and then to the Chattahoochee River.

The Chattahoochee River Line, designed by artillerist Francis Shoup, featured infantry forts alternating with artillery redans and interlocking fields of fire. With the river at his back, Johnston reasoned that the field fortifications would buy him enough time to retreat across it toward Atlanta should the Federals be foolish enough to hazard another frontal assault. Ferguson's and the rest of the AOT occupied the River Line until July 9th, with daily artillery duels. The company lost two men to capture between July 3rd and 5th. Sherman, smarting from his Kennesaw Mountain mistake, threatened the AOT with a river crossing on Johnston's right flank, forcing him to abandon Shoup's line and retreat across the river.

Taking a new line to the south of Peachtree Creek, just a few miles north of Atlanta, on July 17th Johnston was replaced by Gen. Hood, causing despair among the enlisted men. Sam Watkins said that he saw "thousands of men cry like babies–regular, old-fashioned boohoo, boohoo, boohoo."[24]

On July 20th the Federals crossed the creek, and the Confederates attacked, but failed to stop Thomas's Army of the Cumberland. Ferguson's was with Hardee on the right of the line, where the Confederates suffered severely, unable to drive the enemy back across the creek.

By July 21st the battery was with the "artillery, under the command of Brig. Gen. Shoup, massed on the extreme right" of the works constructed to defend the city.[25]

Hood withdrew into the defenses of Atlanta. Gen. Walker, with whom Ferguson's had been so long associated, was killed during the battle of Atlanta on the 22nd when Hardee and Cheatham converged on the Federals near Bald Hill (Fig. 5).

The siege of Atlanta was marked by artillery barrages from Federal siege guns; 5,000 shells were lobbed into Atlanta by Sherman, aiming to destroy the city.[26] The Confederates hunkered down, responding to targets of opportunity along the lines. Ferguson's was resupplied from the Atlanta stores on July 27th and twice in August. On August 31st and Sept. 1st, Hardee's Corps and Ferguson's were part of the forces dispatched south of Atlanta to fight the battles of Jonesborough (where at least one Ferguson's soldier was killed) and Lovejoy Station. During the night of Sept. 1st, the Army of Tennessee evacuated Atlanta, leaving its wreckage to Sherman.[27]

The company suffered casualties August 15th and September 7th. They lost their ammunition wagon, driven by private Spears, in a Federal cavalry charge on the 31st. By the conclusion of the Atlanta campaign, casualties in the three batteries of Martin's Battalion from May to September, 1864 were nine killed, fifty wounded, and eight missing. Of those, two killed, ten wounded and seven missing were from Ferguson's Artillery Co.[28]

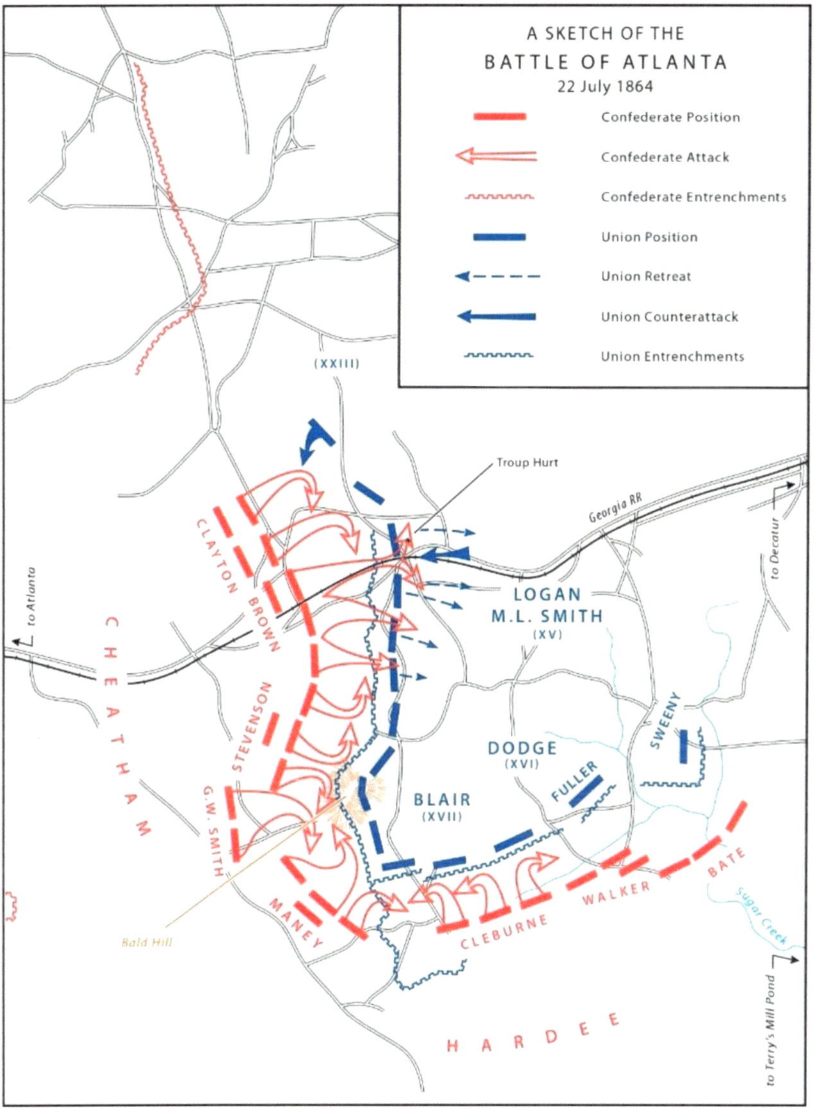

Fig 5. J. Britt McCarley, *The Atlanta and Savannah Campaigns, 1864.* (Washington, D.C.: US Army Center of Military History, 2014)

11. Back to Tennessee

At the end of the Atlanta Campaign, the AOT was at Palmetto, GA, and on Sept. 29th began a 500 mile march which would end at Franklin, TN.[1] It headed north, skirting to the west of Atlanta, marching along a route about halfway between the W&A RR and the GA/AL border, until it reached and captured Dalton. At Mill Creek Gap, Ferguson's was deployed against a Federal block house, but a sister battery did all of the firing, forcing the fortification's surrender.[2] The army departed from there on Oct. 14th, headed southwest, crossing into AL on the 18th on the way to Gadsden, where they drew supplies Oct. 21st. It then went northwest, and crossed the Tennessee River between Tuscumbia and Florence.[3] Sometime between Aug. 31st and Sept. 20th, 1864, the company had left Martin's Battalion of artillery, which had gone to Macon, and Captain Beauregard requested that it be transferred to Major Robert Cobb's Battalion, in Hardee's Corps, with Mebane's (TN) battery and the 5th Co. of the Washington (LA) Artillery.[4] Hardee resigned, and his corps was given to Cheatham. At the AOT artillery HQ in Florence, on Nov. 14th, Cobb's Battalion was assigned to Gen. William B. Bate's division, in Cheatham's corps. Cobb's had twelve guns, 13 officers and 409 men aggregate, but only 227 men present for active duty. There seems to have been some dispute in the long arm as to who was in charge of the battalions: "By this assignment it is not designed or intended to bring in conflict with the division commanders any authority or interest in the artillery heretofore vested in the chiefs of artillery of the corps, but to secure a hearty cooperation in everything that tends to promote the strength and efficiency of the artillery arm of the service. All orders and directions to the battalions from the division commanders will be promptly and strictly obeyed."[5]

Bate's division was at Tuscumbia for two weeks, because the animals were "fagged down and many rendered utterly useless," reflecting the rigors of the march from the Atlanta battles. An additional 360 artillery horses were needed "to supply the places of those broken down and that could not move." Mules were used to replace 240 of those horses in the artillery teams. Baggage wagons were to be converted to carry ordnance and commissary stores, and things were so confused that "it would take a first rate Philadelphia lawyer to sift out this conglomeration of trains."[6] On Nov. 13th Bate marched north after crossing the Tennessee R., to Waynesborough, and then east through Mount Pleasant to near Columbia on Nov. 26th. Captain Beauregard was in command of Cobb's Battalion, on account of Major Cobb being in the hospital.[7] Cobb was still indisposed until at least Dec. 21st, around which time Beauregard returned command of the battalion to him.[8]

On Nov. 29th Bate's division fought at Spring Hill, but it is unlikely that Ferguson's was in action, as the AOT had moved there as quickly as possible, in "light marching order with only a battery" to each corps (which battery went with Cheatham's corps is not stated in the OR).[9] The next day Bate's division was engaged at the battle of Franklin, where it suffered 319 casualties. One unnamed battery accompanied Bate to the front line and dueled with an opposing Federal battery,[10] but most of the southern artillery, likely because of the poor condition of its horses, and because the AOT pursued the Federals from Spring Hill "as fast as possible" on the 30th, did not come up in time for the battle, including Ferguson's. The artillery arrived late, wasn't employed so as to prevent civilian casualties in the town, and was massed for deployment the next morning, by which time the Federals had evacuated toward Nashville.[11] Private Byron Bowers of FAC recounted his memories of Franklin:

> From what I had heard... the slaughter was great and at dawn of day I hit the pike road to go and see for myself. I had gone but a few hundred yards when I came to a dead Federal picket. He was a little round-faced boy, shot in the thigh and bled to death...He had corded his thigh with his own suspender. As I went on up the pike the dead pickets lay thicker...I (walked)... on to about fifty to one hundred yards of the works...I never had seen Confederate carnage so thick...Standing at the gap in the breast works, where the turnpike road passed through, I think there must have been two thousand dead Confederates in sight.
>
> The dead, cold and stiff bodies were laying in every conceivable posture, all with ghastly faces and glassy eyes. Some lay with faces up and some with faces down. Some in a sitting attitude, braced with the dead bodies of their comrades. Some lay with two or three bodies on them. Sometimes you could see a company commander lying with sword in one hand and hat in the other. Sometimes you could see a man who had a heavy martial frown; then again you could see others who wore a pleasing smile...
>
> At the gap in the works where the pike road went through...were lying a Confederate and Federal soldier, both with bayonets sent through their bodies. It was plain to see that they were each other's victims; they both had hands on their guns. Close by a Georgia colonel and a Federal major lay, their positions indicating that they had slain each other with pistols...
>
> When I passed through the gap and looked down (the pike)...another scene of carnage came to view...Streets, gutters, sideway, doorstones, and porticoes were covered with dead men in blue."[12]

Ferguson's marched with Cheatham's corps to Nashville without Bate's division, diverted to Murfreesboro with one battery of Cobb's

Battalion (Slocomb's 5th Co. Washington Artillery). Bate returned to Nashville and rejoined Cheatham's corps and Ferguson's on the 10th or 11.th On the 15th of Dec., his men, many shoeless and feet bleeding on the icy ground, were facing the Federals on the extreme right of the Confederate lines "contiguous to and on the left of the Nolensville turnpike, at a point known as Rain's Hill."[13] (Fig. 1). The remaining Napoleons of Slocomb's battery (two having been left behind in the fighting at Murfreesboro) were positioned at the top of Rain's Hill by Captain Beauregard, who deployed Ferguson's and the rest of Cobb's battalion in this location.[14]

Bate's artillery was attacked here on the 15th by elements of Major Gen. James B. Steedman's division of U.S. Colored Troops. The 14th Colored Infantry led as a skirmish line, followed by the 17th and 44th U.S.C.I., supported by additional regiments and a section of the 20th IN battery. The 3,200 men of Col. Thomas J. Morgan's First Colored Brigade stepped off early, just after a dense fog lifted and the Federal artillery opened. They formed to the right of and parallel to the Murfreesboro Pike, and advanced toward the railroad. The skirmishers received a severe fire, and all troops were "met by a destructive fire at short range from a battery planted on the opposite side of a deep cut made by rail road." Col. Grosvenor reinforced the attack with a brigade of white troops, but all the Federals were "compelled to withdraw...from the range of the artillery."[15] The colonel of the 17th stated that "we were soon obliged to fall back, which was done in rather a disorderly manner... as soon as we were out of range of the enemy's canister we reformed..."[16]

The Confederate perspective of the attack was related by William B. Steele, of Bledsoe's (Mo) battery in Hotchkiss's battalion, fighting next to Ferguson's battery.

> "We were at once ready for action, and Captain Bledsoe gave orders to hold fire - not to fire until the enemy were within twenty paces. Captain B. had his own battery, besides Capt. Goldthwaite's (AL) and Capt. Beauregard's, making twelve Napoleon guns. Our guns were doubled charged with canister, awaiting the near approach of the enemy.

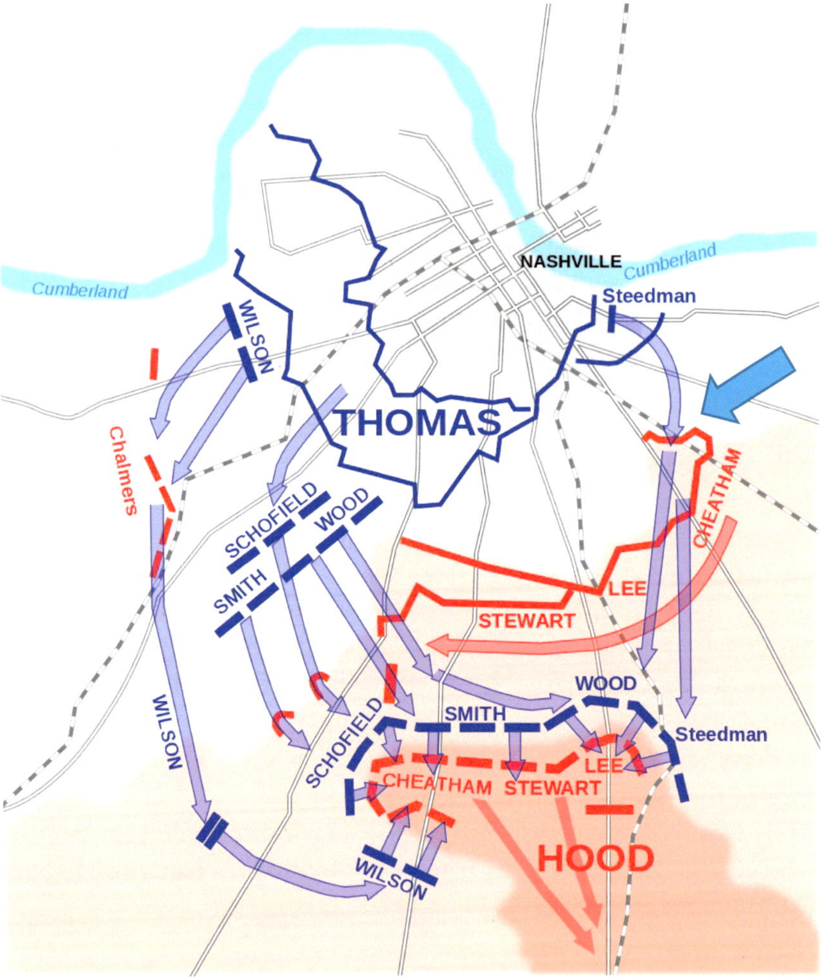

Fig.1. The Battle of Nashville, December 15-16, 1964. Arrow shows Ferguson's with Cheatham 15 December. Wikimedia Commons contributors, "File:Nashville.svg," *Wikimedia Commons*

> Soon we discovered a line of battle – colored troops advancing on us through a blue grass pasture, and behind them a line of white soldiers. We held fire until they were close, when it seemed that every gun was fired at the same time, which created great confusion and panic with the enemy. We fired as fast as we could; the enemy were fleeing in the greatest disorder; we kept up the fire until they were out of sight. We found the field strewn with dead and wounded in our front; one of our men counted sixteen Federal soldiers touching each other, so close were the dead lying."[17]

The Federal attack repulsed by the Confederate right was a feint to divert attention from the main thrust against Hood's left. Cheatham's corps was shifted, with Bate, to the left of the line late on the 15th, in response to the Federal's main attacks. The thawing ground and the difficult terrain forced the artillery to be left behind in their positions near the Nolensville Pike. Hood was forced to retire two miles to a new line of fortifications dug that night.

At daylight on the 16th Cheatham's artillery was retrieved from the right to join Bate's division on the left of the Confederate line. Bate "ordered Captain Beauregard to send a section of howitzers and place them upon a small plateau making out from the declivity of the hill just in rear of Finley's brigade, from which they could sweep the front of my right and the entire line of General Walthall."[18] (Figs. 2 and 3). These guns may have been from Mebane's (TN) battery which had been all 12 pounder howitzers throughout the Atlanta campaign. The balance of Cobb's artillery battalion was either upon or just to the east of what came to be called Shy's Hill. A cannoneer of Slocomb's battery remembered Beauregard's joining his behind the angle formed by Ector and Bate, at Shy's Hill. Bate was battered by musketry and artillery all day, eventually with Federal batteries firing from his left into his rear ranks. The Federal artillery was accurate and effective, battering the hastily built Confederate fortifications.

Captain Beauregard observed a Federal shell approaching and warned a member of Slocomb's battery to duck, the round killing a

Fig. 2 Battle of Nashville (15-16 December 1864). Left arrow is 1st position on Shy's Hill, Dec. 16; right arrow is 2nd position on Franklin Pike (FAC is Ferguson's Artillery) *Atlas to Accompany the Official Records of the Union and Confederate Armies.* Plate 72, 2.

battery horse just behind the lucky man. Reloading the guns was too dangerous in the open, so the crews withdrew them by the prolongs into the fortifications after each discharge.[19] Bate, "anticipating a disaster" as his men fell from fire on three sides and the Federal infantry and dismounted cavalry moved on Shy's Hill, "ordered Captain Beauregard, who commanded my artillery, to move his battalion back to the Franklin Turnpike, as the enemy already had the Granny White pike in our rear, which was my channel for escape."[20]

The Federal advance crushed Col. Shy's command and Bate's extreme left, the 37th GA. On his right, the 29th and 30th GA, under Col. Mitchell of Jackson's brigade, fought until completely surrounded. Here, recounted Bate of Ferguson's: "the section of artillery under Lieutenant Alston was turned on the enemy and fired after being completely flanked and our lines vanishing. I mention it in compliment to this gallant lieutenant." It is evident that despite the order for Beauregard to retire his artillery battalion toward the Franklin Pike, Lt. Alston's section of the company remained in the vicinity of Shy's Hill. The other two Ferguson guns may well have remained there, as well. Bate's division completely gave way, precipitating the retreat of the remainder of the Confederate line. His after-action report lauded Ferguson's commander: "Captain Beauregard, commanding my artillery, showed merit beyond his years, [and] managed the battalion not only to my satisfaction, but to the good of the service and his own credit."[21]

The details of the capture of the battery's four Napoleons are not available, but it seems that when their infantry support fled, the gun crews were forced to leave the pieces. Col. Thomas J. Kinney of the 119th IL INF stated that six guns were captured by his regiment in the action at Shy's Hill.

"After taking the works we advanced through the timber to an open field, and across that to the foot of the mountain, the enemy fleeing before us as fast as possible.

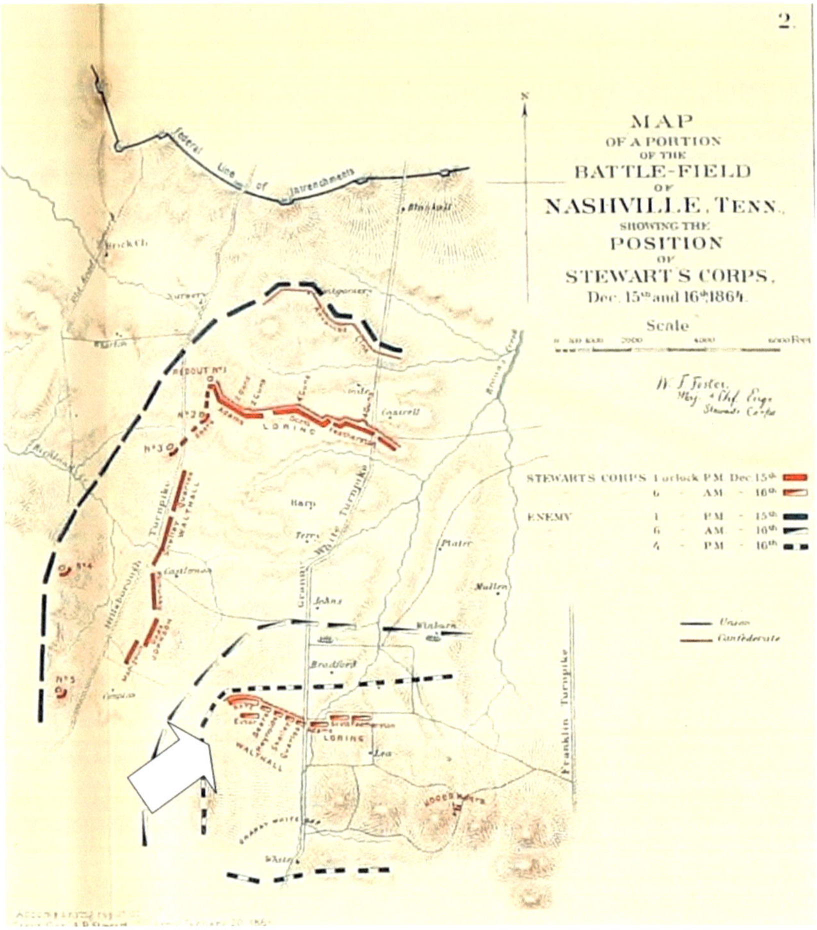

Fig. 3 Arrow on Shy's Hill, Bate's Division and Ferguson's on the Confederate left, Dec. 16th. *Atlas to Accompany the Official Records of the Union and Confederate Armies.* Plate 73, 2.

Here we captured two Napoleon guns and many prisoners. I then...moved by the right flank along the road some distance, and then again by the left flank...and here I captured one 10 pounder Parrott gun, one Rodman, and two more Napoleons. The last four guns captured were the battery of Captain Beauregard, son of General Beauregard, as I was informed by a member of the battery captured near it."[22]

This information does not agree with the fact that Ferguson's had four Napoleons, and no rifled guns, according to the returns of the AOT artillery before the beginning of the Atlanta campaign (the "Rodman" was probably a three inch ordnance rifle, misnamed by infantry who didn't know better). Beauregard himself remembered that Ferguson's had "howitzers" at Nashville, but he could have been referring to Napoleons, also known as "gun howitzers" due to their hybrid design. The fleeing units were all mixed up, and there was ample reason for both Confederates and Federals to be misinformed and confused. It is possible that the first two Napoleons captured in the rear of Shy's Hill were a section of Ferguson's, as were the last two captured farther in the rear. It is just as possible that they were guns belonging to a different battery in Cobb's Battalion. In any event, Beauregard's Battery lost all of its guns somewhere between Shy's Hill and the Franklin Pike. Some men were captured at the guns, and others somewhat later as the Federals mopped up.

Ferguson's had nine men captured, and had lost its guns for the second time. In Cobb's battalion, only Mebane's (Phillips') battery had any guns remaining. Slocomb's Battery dumped its ammunition into a creek and burned its caissons and limbers to keep them from the pursuing Federals.[23] The AOT marched south in a fighting retreat, and left the Federals behind as it crossed the Tennessee River on Christmas Day. Cheatham's corps marched to Corinth, then proceeded by rail and foot to Tupelo, MS, where they camped from Jan. 14th to 17th, 1865.[24]

Captain Beauregard's recollections of Nashville appear in a letter he wrote to Lt. James A. Chalaron of the 5th Co., Washington Artillery, apparently in response to a request from Chalaron.[25] This is the only known written description of battle by any of Ferguson's officers, and is presented here in its entirety, transcribed by the author without alteration of grammar or punctuation.

Letter from René Beauregard to J.A. Chalaron, 4 October, 1875.
A brief history of Ferguson's battery with reminiscences of the Tennessee Campaign.

Ferguson's battery, recruited partly in Louisiana, mostly in South Carolina, was enrolled among the regular batteries of the latter State, and was, with Gist's brigade and other troops from So. Ca. sent to reinforce Gen. Johnston on his assuming command at Jackson, Mississippi. At the close of this campaign and a few days preceding the battle of Chicamauga, in which, however, it did not participate, it was, with its brigade (Gist's) dispatched to the Tennessee army of which it formed an integral part till the capitulation at Greensboro.

At the opening of the Dalton campaign, the vacant captaincy of Ferguson's battery (Capt. Ferguson having been severely wounded at Jackson) was filled by the appointment thereto of the writer, but the company, in accordance to the new artillery organization introduced at that time into the Western army, had been previously attached to Martin's battalion of artillery. However, after the battle of Jonesboro, and shortly before the march of our army into Tennessee, Ferguson's battery was on its Captain's (R.T.B.) application, now transferred to Cobb's battalion of artillery, then consisting of two batteries, one of which was the 5th Co. of the Washington Artillery.

This battery was by no means unknown to the newcomer, that had heard frequently the report of its guns and (if memory be correct here) of the colors taken, on one occasion, by its officers and cannoneers, after the repulse of one of the Enemy's army corps at Jackson Miss.

It had also been thrown into close proximity to it at Pine Mountain, which post could have been made a "sore point" for the enemy had not Gen. Bates, in his usual flighty style commanded, countermanded and recountermanded the firing of the artillery, which, from the cause, was condemned to remain such passive spectators of the gradual advance of the enemy's lines and other fieldworks, that the post was soon so thoroughly under the control of their guns and sharpshooters, that to have held it longer would have been sheer folly.

It was there one evening, shortly after the sharpshooting at dusk had ceased or nearly so, that the writer saw pass litters bearing off three men apparently dead. Inquiring who they were he was informed that they were members of the Washington Artillery, and had all been struck by the same Minié ball. One of the men having been killed outright.

Although 'tis often said that "familiarity breeds contempt" hence that "no man is great for his valet de chambre" the dictum is as often fallacious. Ferguson's battery by its daily contact and daily knowledge of the 5th Co, only learned to appreciate its military worth and its constancy under the trying events of the Tennessee Campaign.

It is doubtful whether any of those present will ever forget the first day of our painful march forward, of the damp and penetrating cold of the banks of the Tennessee River, of the miry mud of the roads, in which the artillery sank up to the axles, guns, limbers and caissons strung along for miles, which neither officers or men - tho' applying their shoulders to the wheels - could move for hours together.

Who can forget the benumbing frostiness of the next day. The writer remembers well asking the 1st Lt of Slocumb's battery if the "icicles, hanging to his moustache, were heavy and how long he had them on?"

On the march the batteries of the battalion were often separated in accordance with special orders transferring them to different parts of the Army. Thus it was on arrival into camp at Columbia, Slocum's battery having been sent forward, the same evening, to some portion of the line and being among the first to cross the river in pursuit of the enemy; but unfortunately too late, from delays in laying the Pontoon bridge, to aid our troops in their first attacks on Franklin.

It was around Nashville only, that the Battalion- since leaving Florence- formed its first regular camp, in which afterwards remained in accord with the masterly status quo of the Army. The third battery (5th Co) having been detached under that worthy commander Gen.

Bates on an expedition against Murfreesboro. This 3rd Co. was however back in time for the battle of Nashville.

On the first day of the battle, it was posted (if memory be correct here) on the extreme right of our lines, where it met and repulsed with slaughter the first negro charge of the campaign.

It was on the evening of this day that our Division was transferred from right to left, and that Major Cobb, unwell for several days previous, was compelled to give up the command of his battalion. This command immediately devolved on the senior Captain, who proved to be the Writer. At first he declined, waving all seniority, but on the persistent refusal of the officers next in grade, he was thus forced to accept it on the following morning, a few moments after the whole of the infantry of our division had moved into line and that the prelude to the battle – by a few artillery shots – had begun.

At about this time, orders for howitzers were sent from Division Hd. Qrs. and Ferguson's battery was accordingly posted, by sections, on the right and left slopes of a very high hill forming part of a range which described a horse shoe, at the extremity of which ran transversely a Turnpike (name not remembered) leading towards Nashville, but feebly guarded by a few vedettes of our extreme left.

The other two batteries of the Battalion were by instruction held in reserve.

As the day wore on thro.' the marches and countermarches of the troops; the attacks and feints of the enemy; some Federal cavalry suddenly charged, at full gallop, down the Turnpike above mentioned. Gen. Cheatam, passing near at the time, ordered the Battalion to move in that direction which it did immediately, but this cavalry did not pause and was as quick to disappear.

On reaching the designated point nothing was seen but a few knapsacks dropped on the road, which some of the vedettes were gladly picking up.

It was here a moment afterwards, whilst grouped on horseback and awaiting events that some of the officers of the Battalion escaped certain destruction. A half spent shell or spherical case fired from some distant gun, plunged passed, with burning fuse, where they were standing, into a puddle of water by the roadside. It was fortunate that Lt. Chalaron suffered not from the failing of near sightedness, for it was by his timely warning that the shell fell harmless.

From this incident onwards the battle waxed hot and fierce. Finally a courier from division Hd. Qrs. was dispatched for the Battalion Commander.

Gen. Bates with some of his staff occupied the declivity of the Hill where the Howitzers of Ferguson's battery had been posted in the morning and were then firing without intermission, for the position was being violently attacked and stormed by infantry, supported in the intervals of their repulse, by field pieces and the heavy ordnance of the River gunboats. Where is the rest of the Battalion Captain asked Genl. Bates?

The place being mentioned, move it, he replied, to a safe distance in the rear as near the Turnpike as you can for things look badly, very badly.

The Writer rode back, but the Battalion, by advice of Lt. Chalaron, had already been moved to a convenient distance. On reaching it, he found the half starved horses feasting on some newly discovered hay and fodder, and was himself preparing to dismount, when suddenly cheers after cheers, followed by volleys of musketry were heard in quick succession. As if by magic, the woods seemed alive with disbanded soldiers flying in every direction whilst guns and limbers caissons and waggons springing seemingly into life at every step were dashing madly against stumps and trees and against each other.

The panic was frightful, the confusion beyond description.

When the Battalion had reached the Pulaski Turnpike (its route of retreat) it was but a dislocated body, the shadow of itself.

Ferguson's battery had saved but one or two limbers and two or three caissons. As to the rest of the other batteries they were equally shattered, but at this distance of time and without data, it would be difficult to state their condition with correctness.

Our discouraging retreat was now begun, and similarly to the opening of the campaign the cold, wet and cloudy nights of its phase added their gloom to the heaviness of our hearts. Frequently, the Enemy's cavalry would dash in among our disorganized rear guards and either capture or destroy the most of them. Resistance was apparently deemed futile by most of the troops which followed along in a mob-like confusion; thus, Pulaski was reached. Around this place we encamped one night after crossing dejectedly its swollen stream, in the muddy banks of which all vehicles would sink to the hubs. Pulaski however, was to prove our first port of refuge, for the intense cold of the following day having frozen hard the miry roads of our travel, the Army resumed its march in better spirits, and finally reached the Tennessee River as the weather had again begun to be wet and cloudy.

The Tennessee was crossed by Cobb's battalion on Christmas.

The Writer, who at Pulaski had given up the command of the Battalion, remembers well as one of the "signs of the times" the Christmas greeting of A. Rost of the 5th Co., just as the batteries were moving from Camp to the Pontoon bridge: Captain I cannot wish you the compliments of the season."

It was across the Tennessee River that Ferguson's battery parted company with the Washington Artillery. The latter (if memory be correct) to garrison and defend Mobile; the former to end its career at Salisbury N.C. after its fight under Gen. Gardner (or Gardiner) with Stoneman's cavalry.

R.T.
Beauregard

12. The Carolinas Campaign

The shattered long arm of the AOT was redistributed. Some battalions were sent to the defense of Mobile, AL, and some to Columbus, GA to defend against Sherman's incursion into the Carolinas. Ferguson's would have gone to Mobile, save the intervention of Gen. Beauregard. Captain William L. Ritter of the Third Maryland artillery recalled that "General Beauregard made a request of General Hood, to send his son's battery, with the first battalion of artillery that was sent to South Carolina. Johnston's battalion being the first ordered there, Captain Beauregard's battery was sent with it, instead of the Third Maryland, which was transferred to Cobb's Battalion, Smith's Regiment of artillery."[1] Maj. J.W. Johnston's battalion, also known as Stevenson's, was comprised of Corput's (GA), Marshall's (TN) and Beauregard's (SC) batteries.

Under the command of the senior captain Marshall, Johnston's battalion left Columbus, MS in early February for what would prove to be a march of two months. The men were paid up to a year of past wages in Columbus, money which would prove handy given the inflated prices encountered along the way (One supper in an Augusta restaurant cost fifty-two dollars for a chicken, side dishes and coffee). Government supply depots were rarely stocked, so the battalion was given "carte blanche" to impress civilian supplies, in exchange for receipts which would no doubt prove worthless with the imminent collapse of the Confederate government. The gun tubes were dismounted and sent by rail to Macon, GA, where the remainder of the battalion would meet them by an overland march. By the time the men arrived at Columbus, GA, they had to rest the horses for five or six days. At Macon, they needed another week of rest. The guns were remounted, and by the time the battalion reached Milledgeville, GA, the horses were so exhausted that the guns were again dismounted and sent by rail to Augusta, where all crossed into South Carolina and camped for three weeks. The battalion moved out in mid-March, through South Carolina, heading through Edgefield, Newberry, and Chester before arriving in Charlotte, NC. On Apr. 3rd, 1865 the battalion arrived in Salisbury, NC and camped for a few days. On the 5th, Johnston's battalion was ordered on a march toward Yorkville (today's York) SC to pasture the horses. Their mission to strengthen the stock was interrupted by events unfolding in Salisbury, and they marched back to arrive there on April 11th.[2,3]

Siting the pieces about 11 pm, in positions selected by a major of engineers from the ANV, Johnston's Battalion found itself covering a four and a half mile front, running south to north from its left flank, and curving north of the town to run west to east to the right flank. The line was up to two and a half miles west of Salisbury, where three roads converged from the west. In the center was Ferguson's battery, on the left was Marshall's, and Corput's battery was on the right (Fig 1). The battalion was sited behind Grant's Creek, running in a steep ravine, with the land beyond the creek timbered, although more open on the left of the line (Fig. 2 and 3). The only infantry support available were about fifty convalescent soldiers, and perhaps 150 galvanized Yankees from the Salisbury prison, liberated upon an oath to fight for the Confederacy.

Early the next morning, two thousand Federal cavalrymen under the command of Gen. Stoneman advanced south toward Salisbury down today's W. Innes St. and Old Mocksville Rd. Those roads' bridges over Grant's Creek north of town were unusable, the Confederates having removed the timber decking. Stoneman sent one hundred mounted men of the 11th KY, those of the 8th TN and the 13th TN, and a force of dismounted men to cross Grant's Creek at three points upstream of the W. Innes St. Bridge. While those units were attacking the Confederates from the west, the brigades of Miller and Brown would replace the bridge decks and advance south on the town.[4]

Before sunrise, the Federal advances from the west drove in the Confederate pickets, and the three batteries of Johnston's Battalion opened fire on the Federal cavalrymen all along their four mile front. As the sun rose, the artillerymen commenced a one and a half hour sustained fire directed at visible cavalrymen and the tree line which hid them, "with a most lavish expenditure of ammunition." This was interrupted twice on the extreme left, where Marshall's two sections, flanking the railroad running west out of town, stopped firing to let two trains depart Salisbury, the second of which was commandeered and burned by the Federals at the Grant's Creek crossing, after its passengers fled back into town. Among these were the wife and daughters of Gen. Polk, who the men of Ferguson's had seen killed at Pine Mountain, GA; his uniform, sword and personal papers were left on the train and captured by the enemy.[5]

Beauregard and Corput withdrew their batteries closer to town, and Marshall retired his pieces to their line about one half mile from Salisbury. Soon after, Federal cavalrymen under Major Keogh of Stoneman's staff swarmed over Marshall's pieces from the rear, having flanked the Confederate left. The same fate befell Ferguson's and Corput's batteries. The infantry support was token, as only a few of the "galvanized" Irish fought well, the majority abandoning the Confederate line to return to the Federal side once the fighting began.[6] Mounted officers and noncommissioned officers escaped capture, but about half of the battalion's soldiers were rounded up by the Federals. By 9 am it was all over, and any soldiers of Ferguson's who hadn't escaped into the woods were now held in the Salisbury prison recently emptied of Federal captives.[7]

Twenty-four men and all of the guns, caissons, etc. of the battery were captured at Salisbury. The Federals claimed eighteen guns were taken, twelve of which were in Johnston's battalion of three batteries.[8] The twenty-four men joined seven hundred prisoners on a forced

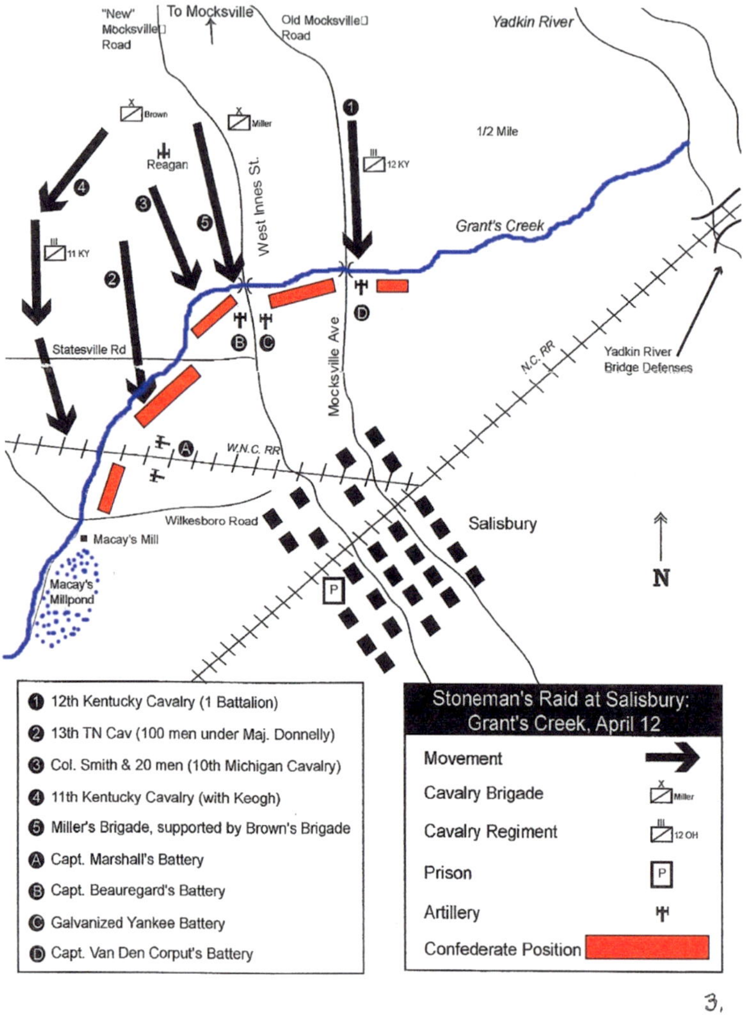

Fig. 1. Stoneman's Raid, https://www.historicalsalisbury.org/education/.

Fig. 2. Grant's Creek, view downstream (north).
Author photo April 21, 2020.

Fig. 3. View west to the tree-lined Grant's Creek.
Author photo April 21, 2020.

march west to Greeneville, TN, and were treated particularly cruelly by the Federal troops. "They were marched rapidly up to the top of the Blue Ridge; several gave out, several who started from Salisbury died. They were all urged forward with threats of death."[9] At Greeneville, they boarded boxcars and travelled by rail through Chattanooga and Nashville on the way to Camp Chase, OH, which they reached on May 4th.[10] The prisoners took the oath of allegiance and left for their southern homes on June 13th. Those of Ferguson's who eluded capture at Salisbury either made their own way home, or were intercepted by Federal forces on the way and made to take the oath before continuing their journey.

Part III: Company Roster

The basis for the compilation of the soldiers who served in Ferguson's Artillery Company is the Combined Service Records (CSR) for the unit. This is conveniently available as a subscription service online at fold3.com. Oddly, some of the men appear under the heading of "Captain Bachman's Co., Artillery (German Light Artillery)," but the majority under the heading of "Capt. Beauregard's Co., Light Artillery (Ferguson's)." Entries for officers can additionally appear under the "Officers" section located in the index between North and South Carolina. René Beauregard, since he was on his father's staff at war's end, is an example of this.

As this compilation is intended to assist the genealogical researcher in a further search for an ancestor, the list has been augmented from multiple sources beyond the CSR. The US census was used to locate each soldier prior to his enlistment. Each man's last known historical record is noted, ultimately his burial place if known. Pension records, obituaries, published works, and other genealogical finding aids filled out the CSR. There are many soldiers in the CSR who remain unidentified in the historical record, and many with further details to discover. The author welcomes inquiries, suggestions and additions from descendants who have access to further information.

The spelling of a soldier's name may be different from how it appears in the CSR, corrections based on records from his post-war life such as marriage and death certificates, the census, or newspaper items.

Conventions of abbreviation in the information are explained in the following example for Private Henry Warren Sargent, arguably the most committed Confederate on the rolls of Ferguson's Artillery Company.

Sargent, Henry Warren 2 Jun 1861 enl. 4th SC INF, Pendleton, Pickens, SC; pvt., 23 y/o; 2 Nov '61 dc'd. for disability GSW head 29 Aug, Manassas, VA; 4 Dec '61 enl. 1st SC CAV, Pickens, SC; 28 Dec '62 dc'd. for disability blind one eye, partial blindness the other; 8 Apr '63 enl. FAC, Pickens, SC; Mar/Apr to May/Jun '63 not yet reported; Nov/Dec '63 pres.; May '64 clothing; 1860 Pickens, SC, Regt. 5, #1044, 22 y/o; b. 1838, d. 1896, bur. Piney Grove Bapt. Cem., Dekalb Co., AL.

Sargent enlisted at Pendleton, Pickens District, SC on June 2nd, 1861 as a 23 year-old private in the 4th SC Infantry. He suffered a disabling gunshot wound to the head in the second battle at Manassas on Aug. 29th, so was discharged on Nov. 2nd. After a period of convalescence, he enlisted in the 1st SC Cavalry at Pickens on Dec. 4th, 1861. His head wound and consequent partial blindness made further cavalry service impossible, and he was again discharged for his disability on Dec. 28th. Ever game, he then enlisted in the artillery, joining Ferguson's at Pickens on Apr. 8th, 1863. In the CSR, presence at a roll call is indicated for the last day of what is usually a two-month period: he had not yet reported to the company and was absent on the last days of April and June of 1863, and was present on Dec. 31st. The CSR usually doesn't indicate where the soldier was at other times during the two months. Sargent received an issue of clothing in May of 1864, and no other information is available for the rest of the war. Records for FAC are incomplete, and the only information after the end of 1863 is clothing issues, and POW records, if applicable. Sargent was in the 1860 census of Pickens, SC, Regiment 5, family #1044, a twenty-two year old. He died in 1896, and is buried in the Piney Grove Baptist Cemetery in Dekalb County, AL, having served the Confederacy in the infantry, cavalry and artillery.

Readers are encouraged to consult a soldier's CSR for additional information, not referenced here for reasons of space. Particularly for the officers, original documents can be illuminating.

Abbreviations

abs	absent for duty
arty	artillery
AWOL	absent without leave
b	born
brig gen	brigadier general
bty	battery
bur	buried
capt	captain
captd	captured
cav	cavalry
cem	cemetery
CM	court martial
co	company
col	colonel
cpl	corporal
d	died
dc'd	discharged
des	deserted
det. serv.	detached service
distr	district
enl	enlisted
FAC	Ferguson's Arty Co
gsw	gunshot wound
hosp	hospital(ized)
inf	infantry
lt	lieutenant
maj	major
MIA	missing in action
NFR	no further record
par	paroled
POW	prisoner of war
pres	present
prom	promoted
pvt	private
regt	regiment
reenl	reenlisted
res	resident
sgt	sergeant
transf	transferred
VA	Veterans Admin.
wd	widowed
wnd	wound(ed)
y/o	years old

Company Roster

* Indicates died in service with Ferguson's Artillery Company

Albreith, John 3 Mar 1862, New Orleans, LA; pvt.; Apr/Jul '62 pres.; Jul/Aug '62 sick, Summerville, SC hosp.; Sep/Oct '62 to Jan/Feb '63 pres.; 29 Dec '63 Dalton, GA, forty days leave, Charleston, for chronic diarrhea, extreme emaciation and debility; clothing issued Apr, Jun, Jul and Aug '64; NFR.

Alston, John Ashe 4 Apr 1862; 3rd lt.; Apr/Jul '62 to Jul/Aug '62 pres.; 28 Oct '62 det. serv.; 29 Dec AWOL; Jan/Feb '63 to May/Jun '63 pres.; Nov/Dec '63 sick furlough; 28 Jul '64 slight neck wnd., Kennesaw Mtn.; 7 Mar '65 abs., commanding bty.; b. 21 Aug 1842; 1860 Georgetown, SC, #809, 17 y/o, student in Europe; d. 6 May 1882, brain tumor; bur. All Saints Episcopal, Georgetown, Pawley's Isl., SC.

***Andrews, James** 25 Feb 1863, Anderson, SC; pvt.; Mar/Apr to May/Jun '63 pres.; 29 Oct '63 pd. $36 for three mos.; d. 9 Dec '63 Empire Hosp., Atlanta, GA; effects claimed by father, John, 3 Oct '64; 1860 Pickens, SC, Regt. 5, #147, 14 y/o.

***Anglesea, Robert** Jan/Feb 1861 enl. 1st Arty.; 4 Feb '62, Charleston, SC transf. to FAC; pvt.; Irish; Apr/Jul to Jul/Aug '62 pres.; 18 Sept '62 jailed for des.; Nov/Dec '62 Charleston, SC jail; CM 7 Feb '63; Mar/Apr to May/Jun '63 jail, 21 May no clothes, blankets; 29 Dec '63 hosp.; d. 23 Jan '64 Medical College Hosp., Atlanta, GA; bur. Oakland Cem., Atlanta.

Ashworth, John J. 19 Feb 1863, Greenville, SC; pvt.; Jan/Feb '63 pres.; Mar/Apr to Nov/Dec '63 AWOL; clothing 21 Mar '64; possibly 1860 Greenville, SC, Caesar's Head, #1039, 24 y/o, and 1870 Transylvania Co., NC, Eastatoey, #37, 33 y/o; NFR.

Ayers, J.H. 1 Jun 1863, Columbia, SC by Maj Melton; pvt.; Nov/Dec '63 hosp. Columbia, SC; 1 Feb '64 General Hosp. No 2; clothing Mar, Apr and Aug '64; NFR.

Bailey, S.T. 16 Nov 1862, Sullivan's Island, SC; pvt.; Nov/Dec '62 to Nov/Dec '63 pres.; clothing Mar, Jun and Aug '64. NFR.

Baldenod, John 3 Mar 1862, New Orleans, LA; pvt.; Apr/Jul to Jul/Aug '62 pres.; 16 Sept jail for des.; Nov/Dec '62 to May/Jun '63 pres.; 25 Nov '63 MIA; 26 Nov '63 captd. Ringgold, GA; POW Rock

Island, IL, 35 y/o; 28 Oct '64 took oath; volunteered and rejected for US service on the frontier. NFR.

Barrett, John 13 Mar 1862, New Orleans, LA; pvt.; Apr/Jul '62 to May/Jun '63, pres., on extra duty as carpenter; 18 Jul '63 captd. Brandon, MS; POW Camp Morton, IN; 10 Mar '64 transf. to Fort Delaware; 30 Oct '64 exchanged. Listed as a conscript. NFR.

Barton, Joberry E. 1 Jul 1862, Pickens, SC; pvt.; Jul/Aug '62 not stated; 26 Jul '62 des. to 49th or 59th GA INF at Jackson, MS (not found in either regt. muster roll) "to be sent for;" 1860 Pickens, SC, Regt. 2, #1106, 20 y/o. Probably same as Joseph Earl Barton, Co. G, 34th GA INF, b. April 1838 (1900 census, Bosque, TX) or 17 Aug 1838, d. 19 Sept 1902, bur. Mosheim, Bosque Co., TX.

Beauregard, Henri T. 2 Jun 1864 appointed cadet, 1st SC INF (Butler's); 15 Feb '65 to report to brother, René T with FAC; 26 Apr '65 paroled, on staff of Gen. Beauregard; 1860 Plaquemines Parish, LA, #278, 15 y/o; b. 11 Apr 1845, d. 5 Feb 1915, bur. St. Bernard Cath. Cem., St. Bernard Parish, LA. Brother of René T.

Beauregard, René T. 12 July 1861 2nd lt., INF; Apr/Jul '62 1st lt., FAC, pres.; Jul/Aug '62 pres.; Sep/Oct to Nov/Dec '62 abs., sick; Jan/Feb to May/Jun '63 pres.; Nov/Dec '63 abs., det. serv.; 23 Apr '64 prom. capt.; 1 May '65 paroled, Greensboro, NC; 1860 Plaquemines Parish, LA, #278, 18 y/o; b. 19 Dec 1843, d. 14 Dec 1910, bur. Metairie Cem., New Orleans, LA. Brother of Henri T.

Beckham, William 11 Apr 1862, Columbia, SC; pvt.; Mar/Apr '63 abs., des. to Preston's SC Arty.; May/Jun '63 abs.; Nov/Dec '63 pres., 25 Nov returned from Preston's, wages held; Mar and Apr '64 clothing; Jul and Aug '64 clothing, Ladies Hosp., Columbus, GA; tentative 1860, Richland, SC, #2, 21 y/o; NFR.

* **Benefield, George W.** 5 Dec 1863, Atlanta, GA; pvt.; d. 28 Dec '63, in camp near Dalton, GA; 24 May '64 father, John, filed deceased soldier's claim; 1860 Heard, GA #483, 15 y/o student; bur. West Hill Cem., Whitfield Co, Dalton, GA. All Benefields are brothers.

* **Benefield, Henry** 5 Dec 1863, Atlanta, GA; pvt.; 29 Dec '63 hosp.; 9 Jan '64 d. Medical College Hosp., Atlanta, GA; bur. mass grave in Oakland Cem., Atlanta; 1860 Heard, GA #483, 30 y/o student. All Benefields are brothers.

* **Benefield, James B.** 5 Dec 1863, Atlanta, GA; pvt.; 29 Dec '63 hosp.; d. in hosp., Atlanta, GA; 24 May, '64 father, John, filed deceased soldier's claim; bur. mass grave, Oakland Cem., Atlanta; 1860 Heard, GA #483, 18 y/o student. All Benefields are brothers.

* **Benefield, John** 5 Dec 1863, Atlanta, GA; pvt.; Nov/Dec '63 pres.; 21 Mar '64 clothing; 17 Dec '64 captd. Franklin, TN; POW Camp Chase, OH; d. and bur. Camp Chase, OH; 1860 Heard, GA #483, 21 y/o student. All Benefields are brothers.

* **Benefield, Joseph J.** 5 Dec 1863, Atlanta, GA; pvt.; 29 Dec '63 hosp.; d. in hosp. Dalton, GA; 24 May, '64 father, John, filed deceased soldier's claim; bur. West Hill Cem., Whitfield Co, Dalton, GA; 1860 Heard, GA #483, 16 y/o student. All Benefields are brothers.

Bennett, John 18 Feb 1862, New Orleans, LA; pvt.; NFR.

Bennett, William F. 2 Mar 1863, Hart Co., GA; pvt.; Mar/Apr to May/Jun '63 pres.; 13 Aug '63 may have enl. 24th GA INF, par. May '65, Hartwell, GA; 1860 Hall, Hart Co., GA, #852, 35 y/o; d. 31 May 1887, bur. Old Canon Bapt. Ch., Bowersville, Hart Co., GA.

Burchfield, Elias Campbell 1 Jul 1862, Pickens, SC; pvt.; 15 Jul '62 des. to 2nd SC INF (not in its roster); Mar/Apr to May/Jun '63 des., "to be sent for;" 25 Nov '63 MIA; 26 Nov '63 captd. Ringgold, GA; POW Rock Island, IL, to 17 Jun '65, 27 y/o; 1860 Pickens, Regt. 2, SC, #1422, 32 y/o; b. 13 Mar 1828, d. 31 Jan 1902, bur. Chauga Cem., Westminster, Oconee Co., SC.

* **Blackston, Thomas C.** 1 Jul 1862, Pickens, SC; pvt.; Jul/Aug '62 to Sep/Oct '62 pres.; 4 Dec '62 d. in camp, Sullivan's Island, SC; 24 April, '63 widow, Dicey, claim for settlement; 1860 Pickens, SC, Regt. 5, #141, 30 y/o, wife and 6 children.

Blanchard, Oscar 10 Mar 1862, New Orleans, LA; pvt.; Apr/Jul '62 to May/Jun '63 pres.; 25 Nov '63 MIA; 27 Nov '63 captd. Graysville, GA; POW Rock Island, IL, to 11 Oct '64, 15 y/o; 1850 could be either St. James, LA #316, 1 y/o or Iberville, LA #352, 3 y/o; NFR.

Bowers, Byron 24 Aug 1861 enl. 24th GA INF, Hart Co., GA; pvt.; Apr '63 transf. to FAC; 19 Apr '63 clothing; Nov/Dec '63 pres.; May and Jul '64 clothing; 1860 Hall, Hart Co., GA, #841, 20 y/o farmhand; b. 21 Jul 1839, d. 21 Jun 1924, bur. Oak View Cem., Canon, Franklin Co., GA.

Brackett, George B. 5 Mar 1862, New Orleans, LA; pvt.; Apr/Jul '62 to Nov/Dec '62 pres.; Jan/Feb to Nov/Dec '63 pres., hospital steward; clothing Jun and Aug '64; 2 Sept '64 to report to Med. Director, Atlanta, as steward; NFR.

Bristow, John M. pvt.; 7 May 1863 des.; 30 Nov '63 reported; May, Jun and Jul '64 clothing; 12 Apr '65 captd. Salisbury, NC; POW Camp Chase, OH to 13 Jun '65; 1860 Marlboro, SC #590, 20 y/o; b. 22 Dec 1835, d. 2 Oct. 1911, bur. Salem Baptist Ch., Bennettsville, Marlboro Co., SC.

Brock, William L. 29 April 1863, Pickens, SC; pvt.; Mar/Apr '63 abs., furloughed; May/Jun '63 pres.; Nov/Dec '63 abs., sick; NFR.

Brooks, John M. 11 Feb 1863, Pickens, SC; pvt.; Jan/Feb to Nov/Dec '63 pres.; Apr '64 clothing; 6 Jun '64 severe hand wnd., Dallas, GA; Jun and Aug '64 clothing; NFR.

Brooks, Taylor 23 Mar 1863, Anderson, SC; pvt.; Mar/Apr to Nov/Dec '63 pres.; Apr, Jun, Jul and Aug '64 clothing; NFR.

Browder, Adam 29 Dec 1861 maybe enl. in 25th SC INF; 18 Jul 1863, FAC, Columbia, SC; pvt.; Nov/Dec '63 pres.; Apr/May '64 clothing; 1860 Williamsburg, SC, #102, 21 y/o; b. 26 Dec 1824, d. 8 Aug 1918, bur. Bethel Baptist, Olanta, Florence Co., SC.

Brown, Charles A. 22 Mar 1862, New Orleans, LA; pvt.; Apr/Jul '62 to Nov/Dec '63 pres., hosp. cook/nurse; Aug '64 clothing; 12 Apr '65 captd. Salisbury, NC; POW Camp Chase, OH to 13 Jun '65, 31 y/o, res. of Orleans Parish, LA; NFR.

Brown, W.J. pvt.; 19 April 1864 clothing; NFR.

Brown, W.O. pvt.; 8 Aug 1864 clothing; NFR.

Brownlee, John W. 9 Mar 1863, Elberton, GA; pvt.; Mar/Apr '63 furloughed; May/Jun '63 abs.; Nov/Dec '63 det. serv.; Mar, cpl., Jul and Aug '64 (pvt.) clothing; 1860 Elbert, GA #893, 35 y/o, 3 children; 3 Jan 1907 GA Indigent Veteran pension application; 1910 Longstreet, Elbert, GA #167, 80 y/o; NFR.

Brownlee, Robert Chester 14 Feb 1863, Abbeville, SC; pvt.; Mar/Apr '63 to Nov/Dec '63 not yet reported for duty; 17 Oct. '64 enl. 2nd SC Rifles, surrendered at Appomattox; 1860 Abbeville, SC

#2126, 16 y/o; b. 1845, d. 1901, bur. Greenville Presb. Ch., Shoals Junction, Greenwood, SC.

Bryson, William 1 Jul 1862, Pickens, SC; pvt.; Jul/Aug '62 pres.; 13 Sep '62 des.; Nov/Dec '62 to May/Jun '63 pres.; 25 Aug Walker's Div. hosp., Lauerdale, MS; Oct '63 furloughed; Nov/Dec '63 abs.; 1860 Pickens, Regt. 5, SC, #95, 22 y/o; 1880 Pickens, Eastatoe, Distr. 132, #14, 51 y/o; NFR.

Buford (Beauford), Moses B. 1 April 1863 enl. 20th SC INF; 28 May '63 dc'd; enl. FAC 1 Sept 1863, Columbia, SC; pvt.; Nov/Dec '63 pres.; 19 Jul '64 Ocmulgee Hosp., Macon, GA; 15 Aug '64 severe hand wnd., Atlanta, GA; 1860 Abbeville, SC #1423, 32 y/o; b. 12 May 1825, d. 8 Nov 1904; bur. Chapel Hill, Hinds Co., MS.

Bull, William Izard, Jr. 26 Jul 1861 appointed asst. surgeon 1st SC (McCreary's); Oct. '61 resigned; 28 Apr '62 enl. Marion Arty.; 21 Mar '63 passed Med. Exam Board, then joined FAC as asst. surgeon; May '65 captd./paroled with Washington Arty. (LA); 1860 Charleston, Ward 2, SC, #395, 22 y/o M.D.; b. 19 Jan 1838, d. 29 Jul 1917, bur. Magnolia Cem., Charleston, SC.

Burdine, James Talley 17 Feb 1863, Pickens, SC; pvt.; Mar/Apr to Nov/Dec '63 pres.; Mar and Jun '64 clothing; 1860 Pickens, SC, Regt. 5, #173, 15 y/o; b. 3 Mar 1854, d. Oct. 1897, bur. Holly Springs Baptist Ch., Pickens, SC.

Burke, David 8 Mar 1862, New Orleans, LA; pvt.; Apr/Jul '62 dc'd. for disability; NFR.

Burke, Thomas W. 23 Jan 1862, Charleston, SC; sgt.; Apr/Jul to Jul/Aug '62 pres.; Sep/Oct '62 demoted to pvt., AWOL; NFR.

Burnette, John 18 Feb 1862, New Orleans, LA; pvt.; 24 Jun '62 dc'd. for disability; NFR.

Burns, John 20 Feb 1862, New Orleans, LA; pvt.; Apr/Jul '62 to Nov/Dec '63 pres.; 9 Mar '64 pd. $24 two mos. service; Apr, Jun and Jul '64 clothing; 28 Sep '64 Bemiss Hosp., Opelika, AL; 57 y/o Irish laborer; NFR.

Burton, J.H. clothing receipts 30 Apr, 7 Jun, 27 Jul, 8 Aug, and 28 Aug '64; NFR.

***Burton, Joseph G.** 31 May 1861 enl. Co. H, 7th GA INF, des. 27 Feb '62; 25 Mar 1863 enl. FAC, Pickens, SC; pvt.; Mar/Apr to

May/Jun '63 pres.; d. 25 Sep '63, Atlanta, GA CSA hosp., bur. Oakland Cem., Atlanta, GA; 11 Mar '64 father, William C., made claim for deceased soldier; 1860 Cobb, GA, Gritter, Marietta P.O., 17 y/o, b. SC.

Burton, Peter MJ 1 Apr 1862, Anderson, SC; pvt.; Apr/Jul '62 pres.; Jul/Aug '62 hosp. Summerville, SC; 1 Sep '62 des.; Sep/Oct '62 Pickens, SC jail; Nov/Dec '62 to Jan/Feb '63 pres.; 28 Feb '63 Capt. Ferguson requests his release from jail after des.; 2 Mar '63 CM; Jun and Aug '64 clothing; 19 Dec '64 captd. Triune, TN; POW Camp Chase, OH to 11 Jun '65, 21 y/o; b. April, 1844, d. 13 Dec 1914, bur. Fairview Community Ch., West Union, Oconee Co., SC; VA headstone installed by author.

Calhoun, Henry Davis lt.; Apr/Jul 1862 recruiting; Jul/Aug '62 not stated; Sep/Oct '62 pres.; Nov/Dec '62 to Mar/Apr '63 det. serv.; May/Jun '63 not stated; Jul to Oct '63 recruiting and equipping the company; Nov/Dec '63 det. serv.; Jun '64 abs., accounted for; Jul '64 pres.; Aug '64 "Col. Smith's command;" Feb and Mar '65 drew pistol cartridges at Macon, GA; 1860 Pickens, SC, Regt. 5, #1250, 21 y/o farmer, b. Pickens; d. 27 Mar '66, bur. Colhoun cemetery in Clemson U. Experimental Forest; VA headstone installed by author.

Cameron, John C. 25 Feb 1862, New Orleans, LA; sgt.; Apr to Aug '62 pres.; 28 Sep '62 Summerville, SC hosp. to Oct; Nov/Dec '62 to May/Jun '63 pres.; 15 Jul '63 wnd., Jackson, MS, to French's Div. Hosp., Lauerdale, MS; 29 Oct '63 pd., Charleston, SC; Nov/Dec '63 in hosp.; 3 Feb '64, pd., disabled; NFR.

Cantrell, John Tyler 18 Jan 1862 enl. 22nd SC INF, dc'd. 7 Apr; 17 Feb 1863 enl. FAC, Pickens, SC; pvt.; Jan/Feb to May/Jun '63 pres.; Nov/Dec, hosp. since 26 Nov; 12 Nov '64 Buckner and Gamble Hosp., Fort Valley, GA; April '65 at Salisbury, NC; 1860 Pickens, SC, Regt. 5, #170, 16 y/o; b. 3 Feb 1844 (pension application), d. 4 Nov 1919, bur. Mountain Grove Bapt. Ch., Pickens, SC.

Carrier, Michael 14 Mar 1862, New Orleans, LA; sgt.; Apr/Jul '62 to Jan/Feb '63 pres.; first week Apr '63 det. serv. arresting deserters; Mar/Apr '63 pres.; May/Jun to Nov/Dec '63 pres.; May, Jun, Jul and Aug '64 clothing; NFR.

Carver, Remus 21 Apr 1863, Pickens, SC; pvt.; Mar/Apr to May/Jun '63 pres.; 1850 Pickens, SC, Western Div., #613, 19 y/o; NFR. Twin of Romus Carver.

Carver, Romus 21 Apr 1863, Pickens, SC; pvt.; Mar/Apr to May/Jun '63 pres.; 15 Jul hosp.; Nov/Dec '63 abs.; 1860 Pickens, SC, Regt. 2, #1297, 28 y/o. NFR. Twin of Remus Carver.

Casey, Thomas Pvt.; Nov/Dec 1862 detailed service; NFR.

* **Cator, William Wesley** 6 Apr 1863, Pickens, SC; pvt.; Mar/Apr to Nov/Dec '63 pres.; 1865 d. in service, wife, Anna Cater, pensioner; 1860 Pickens, SC, Regt. 5, #39, Wesley, 30 y/o.

Chandler, William 14 Feb 1863, Abbeville, SC; pvt.; Jan/Feb to Nov/Dec '63 pres.; Mar, May, Jun and Aug '64 clothing; 12 Apr '65 captd. Salisbury, NC, POW Camp Chase, OH, to 13 Jun '65; 1860 Abbeville, SC, #2078, 13 y/o; b. 26 Sep 1846; SC pension 1906; d. 10 Jun 1927, bur. Rehoboth Bapt. Ch., Greenville, SC.

Chapman, Frederick Dewitt 1 Sep 1863, Columbia, SC; pvt.; Nov/Dec '63 pres.; 21 Mar '64 clothing; 17 Dec '64 des.; 19 Jan '65 CSA Gen. Hosp. 11, Charlotte, NC, typhoid fever; 1860 Darlington, SC #1055, 38 y/o, 4 children; 1880 Darlington, SC, Swift Creek, #288, 58 y/o farmer, eight children; d. before 1900 Darlington, Hartville, Distr. 20, #184, wd. wife, Winnie.

Chapman, John F. 6 Jul 1863, Pickens, SC; pvt; 13 Oct '63 des., Atlanta, GA; AWOL 16 Oct to 1 Dec '63; Nov/Dec '63 pres.; Jan/Feb '64 Blackie Hosp., Madison, GA; NFR.

Chastain, Christopher Columbus Hammond 20 Mar 1862, Pendleton, SC; pvt.; Apr/Jul '62 pres.; Jul/Aug '62 Summerville, SC hosp.; Sep/Oct '62 pres.; 19 Dec sick furlough, Nov/Dec '62 abs.; Jan/Feb to Nov/Dec '63 pres.; Mar, Apr, Jun and Aug '64 clothing; 1860 Anderson, SC, Regt. 42, #183, 16 y/o; b. 22 Dec 1845, d. 14 Nov 1916, bur. First Bapt. Ch., Walhalla, Oconee Co., SC. Brother of William R.

Chastain, L. 21 Mar 1864 clothing; tentative 1850 Anderson, SC, East, #417, Leroy W. Chasteen, 6 y/o; NFR.

* **Chastain, William R.** 28 May 1862, Anderson, SC; pvt.; Jul/Aug '62 Summerville, SC hosp.; Sep/Oct '62 des.; Nov/Dec '62 pres.; 10

Feb '63 d. Camp Helen, Mt. Pleasant, SC; 1850 Anderson, SC, West, #144, W.R., 7 y/o. Brother of Christopher Columbus Hammond.

Clark, James Robert Pvt.; 12 Apr 1865 captd. Salisbury, NC, POW Camp Chase, OH to 13 Jun '65; 1860 Lexington, SC, Saluda, #146, 13 y/o; b. 12 Apr 1847, d. 8 Oct 1910, bur. Liberty U. Method. Ch., Augusta, Richmond Co., GA.

Classen, Charles 15 Mar 1862, New Orleans, LA; pvt.; Apr/Jul to Sep/Oct '62 det. serv. on gunboat; transf. to CS Navy; NFR.

Coffee, Timothy 21 Feb 1862, New Orleans, LA; pvt.; Apr/Jul '62 to May/Jun '63 pres.; 25 Nov '63 MIA; 26 Nov '63 captd. Ringgold, GA, POW Rock Island, IL to 21 Apr '64, 24 y/o; NFR.

Cole, Larkin Anderson 13 Jan 1863, Anderson, SC; pvt.; Jan/Feb to Mar/Apr '63 pres.; 5 Jun sick in hosp.; May/Jun to Nov/Dec '63 pres.; Apr '64 det. with Cpt. Horr; 30 Sep '64 clothing; 19 May '65 captd. Hartwell, GA; 1860 Anderson, SC, Regt. 42, #269, 15 y/o; b. 23 Oct 1844, d. 27 May 1912, bur. Double Springs Bapt. Ch., Anderson, SC.

Coleman, Thomas 12 Mar 1862, New Orleans, LA; pvt.; Apr/Jul '62 to Nov/Dec '63 pres.; Jun, Jul and Aug '64 clothing; NFR.

Collins, Patrick 18 Feb 1862, New Orleans, LA; pvt.; Apr/Jul '62 to May/Jun '63 pres.; 25 Nov '63 MIA; 26 Nov '63 captd. Ringgold, GA, POW Rock Island, IL to 6 Oct '64, released to join US Navy, rendezvous Chicago 22 Jan. 1864; NFR.

Cothran, Joseph (Josiah) N. 6 Apr 1863, Pickens, SC; pvt.; May/Jun to Nov/Dec '63 pres.; 1860 Greenville, SC, Stirling Grove, #1598, 19 y/o; 1880 Pickens, SC, Central, #133, 38 y/o; NFR.

Crocker, James Jefferson 15 Feb 1863, Anderson, SC; pvt.; Mar/Apr '63 sick in camp; May '63 in hosp.; Nov/Dec '63 hosp. since May; wnd. left ankle and right knee (1915 GA pension application); 1860 Hart Co., GA, Reed Creek, #178, 21 y/o; b. Mar 1840, d. 8 Apr 1917, bur. Reed Creek Bapt. Ch., Hart Co., GA, (gravestone says 15th GA INF, but not in that unit's muster rolls). Brother of Lemuel P.

Crocker, Lemuel P. 1862 militia 1115th Distr., Hart Co., GA; enl. FAC 6 Mar 1863, Anderson, SC; pvt.; Mar/Apr '63 sick in camp; May/Jun '63 pres.; May/Dec '63 abs., sick furlough since 17 Sept; Mar

'64 clothing; 11 Apr '64 unfit for field serv., sent to Atlanta; 1860 Hart Co., Ga, Reed Creek, #178, 23 y/o; 1870 Reed Creek, Hart, GA, #899, 30 y/o; d. before 1880; NFR. Brother of James Jefferson.

* **Crookes, William** 22 Feb 1862, New Orleans, LA; pvt.; Apr/Jul '62 to May/Jun '63 pres.; 25 Nov '63 KIA Missionary Ridge, TN.

Dailey, James 19 Feb 1862, New Orleans, LA; pvt.; 4 Jun '62 dc'd for loss of eye and epilepsy; 50 y/o Irish laborer; NFR.

Dangerfield, William 15 Apr 1862, Berkeley Co., SC; pvt.; Apr to Aug '62 pres.; 7 Sep des., Sep/Oct '62 abs.; 31 Jul '63 enl. Co. E, 5th SC CAV; on sick furlough 5 Aug '64 pd. at Charleston, SC for Apr to Oct '62; 1 Dec '64 clothing; 1860 Charleston, St. Johns Berkeley, SC, family #143, 27 y/o; b. 1833, d. 1878, bur. at "Bamboretta," family plot near Moncks Corner, SC; VA headstone installed by author.

David, Joseph A. 17 Dec 1864 captd. Franklin, TN, POW Camp Chase, OH to 11 Jun '65, res. of Abbeville Distr., SC; 1850 Abbeville, SC, #91, 10 y/o; 1900 Leesburg, Lake Co., FL Distr. 72, #260; NFR.

Davis, Frederick Moss 13 Jan 1863, Pickens, SC; pvt.; Jan/Feb to Dec '63 pres.; 7 Jun '64 clothing; 1860 Pickens, SC, Regt. 2, #1535, 15 y/o farmer; b. 14 Oct 1844, d. 23 Sep 1912, bur. Beaverdam Bapt. Ch., Fair Play, Oconee Co., SC; brother of James P.

Davis, James P. 26 Mar 1863, Pickens, SC; pvt.; Mar/Apr '63 pres.; 15 Jun prom. cpl., May/Jun '63 pres., listed as quartermaster sgt.; probably 1860 Pickens, SC, Regt. 2, #1535, 24 y/o, brother of Frederick M.; NFR.

Davis, James 26 Apr 1863, Pickens, SC; pvt.; Mar/Apr '63 pres.; 2 Jun AWOL, May/Jun '63 abs.; 25 Nov '63 MIA; 27 Nov '63 captd. Ringgold, GA, POW Rock Island, IL, res. of Anderson, SC; 1860 Anderson, SC, Regt. 42, #560, 23 y/o, 2 children; tentative 1880 Belton, Anderson Co., SC, #332, 45 y/o; NFR.

Dayton, Robert 21 Mar 1864 clothing; NFR.

Decker, James HC 21 Feb 1862 New Orleans, LA; pvt.; Apr/Jul '62 to May/Jun '63 pres.; 25 Nov '63 MIA; 27 Nov '63 captd. Ringgold, GA, POW Rock Island, IL to 25 Jan '64 when joined US Navy; USN pension application claimed deafness and scalp wnd. at

Ft. Fisher, NC; 1860 New Orleans, LA, Ward 10, #355, 15 y/o; b. 8 Jan 1845, d. 3 Feb 1917, bur. Mobile National Cem., Mobile, AL.

DeLapierre, Jean Francois 20 Feb 1862, New Orleans, LA; pvt.; Apr/Jul '62: 62 y/o Swiss gardener dc'd, overage; NFR.

Dixon, Charles L. 22 Jan 1862, Charleston, SC; pvt.; transf. from 1st SC INF (Butler's); Apr/Jul to Sep/Oct '62 pres.; Nov/Dec '62 Sgt., pres.; Jan/Feb '63 pres.; Mar/Apr '63 pres., CM; Nov/Dec '63 abs., 5 Nov St. Mary's Hosp., Lagrange, GA; 3 Dec '63 clothing; 30 Mar '64 2 mos. furlough for extreme emaciation, debility and chronic bronchitis; NFR.

Dobson, E. 21 Mar 1864 clothing; NFR.

Dobson, William A. 6 Mar 1862, Pickens, SC; pvt.; Apr/Jul to Sep/Oct '62 pres.; Nov/Dec '62 pres., CM 11 Nov; Jan/Feb to Nov/Dec '63 pres.; Jun, Jul and Aug '64 clothing; 16 Dec '64 captd. Nashville, TN, POW Camp Douglas, IL to 14 Mar '65; 24 Mar '65 at Jackson Hosp., Richmond, VA, sick, paroled prisoner; 1860 Pickens, SC, Regt. 5, #1012, 18 y/o; 1910 Eastland, Ranger, TX, Distr. 39, #232; NFR.

Donald, J.R. 14 Feb 1863, Abbeville, SC; pvt.; Jan/Feb to Nov/Dec '63 not reported to company; NFR.

Donnelly, Bernard 19 Feb 1862, New Orleans, LA; pvt.; Apr/Jul to Jul/Aug '62 pres.; 16 Sep jailed, Charleston, for des.; Sep/Oct to May/Jun '63 in jail after CM; 12 May '63 has neither shoes, socks nor coat in jail, not given meat regularly; 11 Nov '63 sentenced to firing squad on 14 Nov; Nov/Dec '63 pres.; Jan/Feb '64 at Blackie Hosp., Madison, GA; NFR.

Dorsey, A.J. 14 Jun 1862, Jackson, NC; pvt.; Mar/Apr to Nov/Dec '63 not yet reported to company; NFR.

Duncan, James William 25 Dec 1863, Marietta, GA; pvt.; Nov/Dec '63 not reported yet; Mar, May, Jun, Jul and Aug '64 clothing; 16 Dec '64 captd. Nashville, TN, POW Camp Douglas, IL to 19 Jun '65, resident of Edgefield Distr., SC; 1860 Pickens, SC, Regt. 2, #774, 22 y/o; b. 1 Mar 1840, d. 18 Feb 1920, bur. First Mt. Moriah Bapt. Cem., Greenwood Co., SC.

Duncan, T.P. 29 Mar 1863, Pickens, SC; pvt.; Mar/Apr to Nov/Dec '63 not yet with company; NFR.

Dunn, Samuel 21 Mar 1862, New Orleans, LA; pvt.; Apr/Jul '62 in Columbia, SC hosp.; Jul/Aug '62 to May/Jun '63 pres.; 27 Aug '63 AWOL; NFR.

Dunn, W.R. 14 Feb 1863, Abbeville, SC; pvt.; Mar/Apr to Nov/Dec '63 never reported to company; NFR.

Dunnahoo, James H. 5 Nov 1861, Pickens, SC; pvt., 2nd Regt SC Rifles; 27 Jun '62 severe hand wnd., Gaine's Mill, VA; 5 Nov '63 enl. FAC in exchange for James B. Hughes; 25 Nov MIA; 26 Nov '63 captd. Ringgold, GA, POW Rock Island, IL to 11 Oct '64; USA enl. rejected by enrolling officer; 1860 Pickens, SC, Regt. 2, #423, 18 y/o; 4 Jun 1897 Colbert, AL pension application, claimed broken right arm/shoulder Malvern Hill, VA; NFR.

Dunnevant, William T. 10 Feb 1862, Huntsville, AL; pvt.; Apr/Jul '62 to May/Jun '63 pres., articifer 1st class/blacksmith, repairing bty.; 17 Jul '63 captd., Jackson, MS, POW Camp Morton, IN to 10 May '65, oath to stay north of Ohio River; b. Oct 1830, TN; 1860 AL, Madison SW #19, 28 y/o, 4 children; 1900 IL, Cook, Chicago Ward 30, #83, 69 y/o; NFR.

Durham, Perry A. 5 Mar 1862, Pickens, SC; pvt.; Apr/Jul '62 sick, Pendleton, SC; Jul/Aug '62 pres.; 13 Sep des., Sep/Oct '62 abs.; Nov/Dec '62 jail, Anderson, SC since 11 Dec; Jan/Feb '63 pres.; Mar/Apr to Nov/Dec '63 AWOL since 29 Mar; deserted, took oath to stay north of Ohio R. Jan. '65; 1860 Pickens, SC, Regt. 5, #985, 17 y/o; 1866 murder conviction, escaped SC Lunatic Asylum; recaptured, sentence commuted 21 Dec., 1868; NFR.

* **Duval, Henry** 3 Mar 1862, New Orleans, LA; pvt.; Apr/Jul '62 pres.; Jul/Aug '62 hosp., Summerville, SC; 5 Sep '62 d. in hosp.; 1860 New Orleans, LA, 3rd Ward, workhouse, line 5, 27 y/o, b. England.

Eads, Samuel H. 11 Feb 1863, Pickens, SC; pvt.; Mar/Apr to Nov/Dec '63 pres.; Apr, Jun, Jul and Aug '64 clothing; 1860 Pickens, SC, Regt. 5, #1075, 14 y/o; b. 1836, d. 1929, bur. Rose Cem., Prairie Grove, Washington, AR.

Eades, Christopher Columbus 10 Feb 1862, Pickens, SC; pvt.; Apr/Jul to Sep/Oct '62 pres.; Nov/Dec '62 abs., sick furlough 19 Dec; Jan/Feb to Nov/Dec '63 pres.; Jun and Sep '64 clothing; 1860

Pickens, SC, Regt. 5, #967, 15 y/o; b. 30 Sep 1845, d. 25 May 1903, bur. Six Mile Bapt. Ch., Pickens, SC.

Edwards, J.W. Jul and Aug 1864 clothing; NFR.

Elbrook (Elbrock), William M. 8 Sep 1861 enl. Co. E, 1st SC Arty, drummer boy, dc'd 15 Aug '62 for injuries from a fall; enl. 2 Sep '62 in Co. A, 20th SC INF, pres. to Feb '63; 1 Apr 1862 enl. FAC, Columbia, SC; pvt.; 27 Oct '63, Gate City Hosp., Atlanta, GA, certified disability, 16 y/o b. Charleston, SC, night blindness and want of stamina, discharged to Walhalla, SC; 1860 Charleston, Christ Church, SC, #85, 11 y/o; b. 1847, d. 14 Mar 1914, bur. Confed. Cem. #2, Chilton Co., AL.

* **Elford, Thomas George** 29 Jan 1862, Charleston, SC; pvt.; Apr/Jul '62 to Mar/Apr '63 pres.; May/Jun '63 Summerville, SC hosp., diarrhea; 1840 Nantucket, MA, 30-40 y/o; b. 23 Dec 1806, Charleston, SC, d. 24 Jul '63, Columbia, SC hosp.

* **Elrod, Elijah J.** 4 Dec 1861 enl. Co. D, 18th SC INF, reenl. 9 Apr '62; cpl; 23 Feb 1863 enl. FAC, Franklin, GA; pvt.; Mar/Apr '63 sick in camp; May/Jun '63 pres.; Nov/Dec '63 abs., 2 Oct sick furlough; Apr '64 clothing; d. 12 May 1864, Dalton, GA; 1860 Anderson, SC, Regt. 42, #884, 36 y/o, five kids and pregnant wife.

Epps, Daniel Edward 19 Jun 1863, Columbia, SC; pvt.; Nov/Dec '63 pres.; Dec '63 Ladies Gen Hosp., Columbia, SC; Mar, May, Jun and Jul '64 clothing; 12 Apr '65 captd. Salisbury, NC, POW Camp Chase, OH to 13 Jun '65; 1860 Clarendon Co., SC, #551, 15 y/o; b. 25 Dec 1844, d. 27 Aug 1917, bur. Midway Presb. Ch., New Zion, Clarendon Co., SC.

Fant, William B. 29 Apr 1863, Pickens, SC; pvt.; Mar/Apr to Nov/Dec '63 abs., never reported to company; 1860 Pickens, SC, Regt. 5, #1262, 29 y/o, 1 child; NFR.

Farrell, Michael 23 Jan '62, Charleston, SC; Sgt.; Apr/Jul to Jul/Aug '62 pres.; Sep/Oct '62 demoted to pvt. for AWOL; Jan/Feb to May/Jun '63 des. to Brook's Arty., "to be sent for"; Oct '63 returned from Brook's "to proper command after deserting same;" Nov/Dec '63 pres.; 17 Jun '64 severe thigh wnd., Pine Mtn., GA; cpl. by 27 Sep '64; Sep, Oct and Nov '64 clothing; 1860 Charleston, SC,

Ward 4, line 26, prisoner, 26, b. Ireland; 1904 Charleston, SC City Directory, p. 353; died 1904-5. NFR.

Ferguson, Thomas Barker 1 Jan 1862, capt.; Apr/Jul '62 to Jan/Feb '63 pres.; Mar/Apr '63 on leave since 28 Apr; May/Jun '63 pres.; 14 Jul '63 chest wnd., Jackson, MS; Nov/Dec '63 abs.; 15 Jun '64 prom. maj; to Jan '65, Charleston Arsenal and Inspector of Arty.; 1860 St. Johns Berkeley Parish, SC, Charleston, #139, 19 y/o; b. Aug 1841, d. 11 Aug 1922, bur. Green Mount Cem., Baltimore, MD.

* **Fortner, Tandy L.** 10 Jun 1862, Pickens, SC; pvt.; enl. in 6th SC CAV before reporting for prior enl. in FAC; Jul/Oct '62 (6th CAV) abs., in Summerville Hosp.; Sep/Oct '62 (FAC) abs., des. 15 Jul; Nov/Dec '62 (FAC), d., Summerville Hosp.; 7 Apr '63, widow, Arminda, filed claim for deceased soldier; 1891 and 1901 pension for Arminda; 1860 Pickens, SC, Regt. 5, #235, 25 y/o, 2 children. Brother of Zion.

* **Fortner, Zion** 1 Jul 1862, Pickens, SC; pvt.; near simultaneous enl. in 6th SC CAV; Jul/Oct '62 (6th CAV) abs. in arrest 19 Oct; Sep/Oct to Nov/Dec '63 (FAC) pres.; Apr, May and Jun '64 clothing; 1 Sep '64 KIA, Jonesboro, GA; 14 Nov '64 widow, Malinda, filed claim for deceased soldier; 1901 widow's pension; 1860 Pickens, SC, Regt. 5, #236, 23 y/o, 1 child. Brother of Tandy L.

Franklin, William Elbert 1 Sep 1863, Edgefield, SC; pvt.; Nov/Dec '63 pres.; Mar, May, Jun, Jul and Aug '64 clothing; 1860 Edgefield, SC #179, 16 y/o farmer; d. 7 Feb 1871, bur. Mt. Cavalry Lutheran Cem., Aiken, SC.

Freeman, H.R. 12 Jul 1863, Millwood, SC; quartermaster sgt.; Nov/Dec '63 abs., recruiting; Jun '64 clothing; NFR.

Freeman, Henry W. 14 Feb 1862, Pickens, SC; pvt.; Apr/Jul '62 pres.; Jul/Aug '62 Summerville, SC hosp.; Sep/Oct '62 pres.; Nov/Dec '62 sick furlough 19 Dec; Jan/Feb to Nov/Dec '63 pres.; Apr '64 clothing; 2 Jul '64 MIA, Marietta, GA; 3 Jul '64 captd. Kennesaw Mtn., GA; 6 Jul '64 US hosp., Chattanooga, TN, acute diarrhea; 19 Sept exchanged; 12 Apr '65 captd. Salisbury, NC, POW Camp Chase, OH to 13 Jun '65; 1850 Pickens, SC, East, #49, 9 y/o; b 23 May 1840, bur. Pleasant Gap Bapt. Ch., Elijay, GA.

Friday, Louis 20 Aug 1863, Edgefield, SC; pvt.; Mar, Apr, Jun, Jul and Aug '64 clothing; 1860 Pickens, SC, Walhalla, #549, 19 y/o clerk; possibly in Siskiyou Co., CA, 1878; NFR.

Frier, J.W. 10 May 1862, Pickens, SC; pvt.; 20 Jun '62 AWOL; 22 Jul returned to camp; 26 Jul des.; NFR.

Garner, John 29 Oct 1862, Sullivan's Island, SC; pvt.; Sep/Oct '62 to Nov/Dec '63 pres.; Jul and Aug '64 clothing; 7 Sep '64 MIA, Jonesboro, GA; NFR.

Garrett, D. 21 Mar 1862, New Orleans, LA; pvt.; Apr/Jul to Sep/Oct '62 pres.; NFR.

Geary, Michael 8 Feb 1862, Charleston, SC; pvt.; Apr/Jul '62 Summerville, SC hosp.; Sep/Oct to Nov/Dec '62 pres.; 14 Feb '63 cpl.; Jan/Feb to May/Jun '63 pres.; Nov/Dec '63 pres., pvt.; 16 May '64 captd. Calhoun, GA; POW Rock Island, IL, left 10 Jun '64, enl. USN; 8 Aug '64 clothing; NFR.

Gibson, Reuben W. 29 Apr 1863, Pickens, SC; pvt.; May/Jun to Nov/Dec '63 pres.; May and Aug '63 clothing; 5 Sep '64 retired from Invalid Corps; 1860 Pickens, SC, Regt. 2, #777, 15 y/o; b. 23 May 1844, d. 27 Sep 1887, bur. S. Union Bapt. Ch., Fair Play, Oconee Co., SC.

Gilliland, David J. 8 Apr 1863, Pickens, SC; pvt.; Mar/Apr '63 abs. on furlough; May/Jun '63 pres.; 1860 Pickens, SC, Regt. 5, #533, 28 y/o, 1 child; 1880 Pickens, SC, Pickens C.H., #162, 52 y/o; NFR.

* **Gilliland, James A.** 1 Jul 1862, Pickens, SC; pvt., 17 y/o; simultaneous enl. 6th SC CAV, arrested 19 Oct, abs. from 6th CAV 31 Oct; (FAC) Sep/Oct '62 pres.; Nov/Dec '62 abs., sick furlough 29 Nov; Jan/Feb to May/Jun '63 pres; hosp. Casswell, GA sometime between Jul/Oct '63; Nov/Dec '63 abs., hosp. 21 Dec; May and 7 Jun '64 clothing; 28 July '64 KIA near Marietta, GA; 14 Nov '64 mother, M.A., applied for deceased soldier claim; 1860 Pickens, SC, Regt. 5, #215, 16 y/o.

Girard, Dominick 21 Mar 1862, New Orleans, LA; pvt.; Jan/Feb to May/Jun '63 pres.; Nov/Dec '63 MIA since 25 Nov '63; NFR.

Gorman, George 18 Feb 1862, Charleston, SC; pvt.; Sep/Oct '62 abs., det. duty on gunboat; NFR.

Gravely, Lontford Ballenger 6 Apr 1863, Pickens, SC; pvt.; Mar/Apr '63 not yet reported; May/Jun '63 pres.; 9 Aug '63 hosp. Lauerdale Springs, MS; Nov/Dec '63 pres.; 2 Apr '64 unfit for field duty, sent to Atlanta; 19 Apr '65 at Salisbury, NC; 14 Apr 1919 pension application, blind, infirm, 91 y/o; 1860 Pickens, SC, Regt. 5, #286, 30 y/o; b. 26 Jun 1828, d. 25 Jan 1922; bur. Porters Chapel UM Ch., Pickens, SC. Brother of R. Wiley.

Gravely, R. Wiley 19 Mar 1863, Pickens, SC; pvt.; May/Jun '63 pres.; Nov/Dec '63 abs., hosp. since 1 Sep; descendant says he died in Richmond in 1865; 1850 Pickens, SC, East, #742, Wiley, 21 y/o. Brother of Lontford B.

Graves, Barnard F. 16 Nov 1862, Sullivan's Island, SC; pvt.; Nov/Dec '62 to May/Jun '63 pres.; 25 Nov MIA; 1 Dec '63 captd. Ringgold, GA, POW Rock Island, IL to 25 Jan '64; enlisted US Navy, transf. to Naval facility 5 Feb '64; NFR.

Greenrod, James T. 12 Feb 1862, Grand Junction, TN or New Orleans, LA; pvt.; Apr/Jul '62 to Sep/Oct pres., cpl.; Nov/Dec '62 pres.; Jan/Feb '63 pres., reduced from cpl. 14 Feb; Mar/Apr '63 sick in camp; May/Jun '63 pres.; 17 Jul '63 captd. Jackson, MS, POW Camp Morton, IN to 20 May '65, said he was conscripted, and took loyalty oath; British stone cutter; stayed in Indianapolis; bur. 1 Nov. 1889, Crown Hill Cem., Indianapolis; VA headstone provided by reenactors of Ferguson's SC Artillery Company.

Griffiths, George 30 Jan 1861 Co B, 1st Arty., 24 y/o; Mar/Apr '61 in jail, Ft. Moultrie, SC; May to Aug '61 pres.; Sep/Oct '61 des.; 6 Feb 1862, FAC, Charleston, SC; pvt.; Apr/Jul '62 prison since 23 Jun; Jul/Aug '62 pres.; Sep/Oct '62 abs., des. 20 Oct; Nov/Dec '62 jailed 2 Nov; Jan/Feb '63 pres.; Mar/Apr '63 solitary confinement since 7 Apr; 25 May/Jun to Nov/Dec '63 pres., jail dc 15 May; 17 May '64 MIA; captd. 18 May '64, Calhoun GA, POW Camp Morton, IN to 21 Aug '64, took oath of US allegiance, musician/bugler; 1860 Colleton, SC, St. Bartholomew's Parish, #757, 23 y/o, one child; NFR.

Hall, L.B. 6 Feb 1862, Anderson, SC; pvt.; 5 Jul '62 des.; was in Co. G of 6th SC CAV, mistakenly claimed by FAC after presenting to its Sgt.; Capt. Ferguson honored the prior claim of the 6th; NFR.

Hamilton, Andy M. 5 Mar 1862, Williamston, SC; pvt., Co. D, Hampton's Legion; 31 May '62 thigh wnd. at Seven Pines, VA; 4 Oct '63 transf. to FAC; Nov/Dec '63 pres.; May and Sep '64 clothing; 1860 Anderson, SC, Regt. 4, #518, 35 y/o; 1886 disability claim, Pickens, SC; NFR.

Harbot, William Jordan 24 Aug 1861 enl. Habersham, GA, Co. K, 24th GA, pvt., pres. through 31 Oct; enl. FAC 22 Apr 1863, Pickens, SC; pvt.; Mar/Apr '63 pres.; May/Jun '63 hosp. 10 May Jackson, MS, 1 Jun gen. hosp., dysentery; 1850 Pickens, SC, West, #617, William Harber, 8 y/o; 1860 Habersham, GA, #606, William Herbot, 19 y/o; NFR.

Harkins, James 18 Mar 1862, Pickens, SC; pvt.; Apr/Jul to Jul/Aug '62 pres.; Sep/Oct to Nov/Dec '62 det. serv.; Jan/Feb '63, 1 Feb cpl.; Mar/Apr to Nov/Dec '63 pres.; Aug '64 clothing; 12 Apr '65 captd. Salisbury, NC, POW Camp Chase, OH to 13 Jun; 1870 Pickens, Easley, SC, #28, 45 y/o; b. 1827 Donegal, Ireland; d. 27 Mar 1912, bur. Bethel Presb. Ch., Walhalla, Oconee Co., SC.

* **Harper, William P.** 27 Jan 1862, Charleston, SC; pvt.; Apr/Jul '62 to May/Jun '63 pres.; 15 Jul hosp.; d. 19 Sep '63 hosp. Point Clear, Baldwin Co., AL.

Haynes, Andrew 24 Mar 1862, Pickens, SC; pvt.; 15 Jun '62 des., Apr/Jul to Sep/Oct '62 AWOL; 28 Aug '64 clothing; 7 Sep '64 missing at Jonesboro, GA; 1860 Anderson, SC, Williamston, #378, 25 y/o; b. 28 Dec 1837, d. 9 Jan 1917, bur. Camp Creek Cem., Lilburn, Gwinnett Co., GA.

Hering, Barry Benson 2 Mar 1863, Hartwell, GA; pvt.; Mar/Apr to Nov/Dec '63 not yet reported to company; 1860 Hart Co., GA, #673, 29 y/o, 4 children; 1900 Hart Co., Alford, Distr. 53, #177, 71 y/o; NFR.

Higgins, Benjamin H. 21 Sep 1863, Columbia, SC; pvt.; Mar, May, Jun and Jul '64 clothing; 15 Aug '64 slight chest wnd., Atlanta, GA; 12 Apr '65 capt. Salisbury, NC, POW Camp Chase, OH to 13 Jun '65, resident of Richland Distr., SC; 1860 Richland, SC #87, 16 y/o; b. 4 Aug 1845, d. 24 Nov 1914, bur. Columbia Friends Meeting Cem., Richland Co., SC.

Hill, Abraham 23 Feb 1863, Franklin, GA; pvt.; Mar/Apr to May/Jun '63 pres.; 23 Jul '63 enl. 2nd Rifles, present to Dec '65 on furlough of indulgence; 1860 Madison, GA, #229, 26 y/o, b. SC; b. Dec 1827, d. 2 Jul 1907, bur. Franklin Springs Baptist, Franklin Co., GA.

* **Hilliard, Francis Marion** 15 Mar 1863, Anderson, SC; pvt.; Mar/Apr to May/Jun '63 pres.; 21 Dec '63 travel from Athens to Hartwell, GA; 23 Dec d. on furlough, White Co., GA; bur Old Canon Bapt. Ch., Bowersville, Hart Co., GA; 1860 Hart, GA, #838, 32 y/o, 3 children. Likely brother of William A.

Hilliard, William A. 2 Mar 1863, Hart Co., GA; pvt.; Mar/Apr to Nov/Dec '63 pres.; Mar, Apr, Jun, Jul and Aug '64 clothing; 1860 Hart, GA, #848, 34 y/o, 5 children; d. 6 June 1906, bur. Kershaw City Cem., Lancaster Co., SC. Likely brother of Frances Marion.

Hincks, Joseph Anatole sgt., Orleans Guards, LA Volunteers; 4 Feb 1862, enl. FAC, pvt., Mobile, AL; Sep/Oct '62 to May/Jun '63, pres., HQ clerk; 20 Jun returned to company; Nov/Dec '63 acting 2nd lt. without rank; 7 May '64 enl. Co. D, Manigault's Bttn. Art'y; clerk at Dept. HQ until May, '65; 1860 New Orleans, LA, Ward 4, #1076, 17 y/o; d. 9 Dec 1923, bur. Metairie Cem., New Orleans, LA.

Holmes, Jackson 20 Aug 1863, Edgefield, SC; pvt.; Nov/Dec '63 pres.; Mar, Jun and Aug '64 clothing; 1860 Edgefield, SC, #177, 33 y/o; b. 28 Jan 1827, d. 17 Sep 1903, bur. Woodland Cem., Xenia Township, Greene, OH; Mexican War veteran, cpl., 2nd OVI.

Howell, Cleburn F. 25 Jul 1863, Columbia, SC; pvt.; Nov/Dec '63 pres.; Mar '64 clothing; 17 Dec '64 captd. near Franklin, TN, POW Camp Douglas, IL to 6 Apr '65, claimed to be conscript, loyal to US, deserted for amnesty, joined Co F, 5th US Volunteers, mustered out 15 Oct '66, 22 y/o, b. Georgetown, SC; NFR.

Hughes, Henry 23 Aug 1863, Pickens, SC; pvt.; Nov/Dec '63 pres.; Jun and Jul '64 clothing; 12 Apr '65 captd. Salisbury, NC, POW Camp Chase, OH to 13 Jun '65; 1860 Pickens, SC, Regt. 2, #1013, 26 y/o; 1900 Pulaski, Oconee Co., SC, #10, 66 y/o; NFR.

Hughes, James B. 6 Apr 1863, Pickens, SC; pvt.; Mar/Apr '63 furlough; May/Jun '63 pres.; 5 Nov '63 transf. to Co. K, 2nd SC Rifles, in exchange for J.H. Dunnahoo; 16 Aug '64 KIA on picket, New Market Hill, VA; 1860 Pickens, SC, Regt. 5, #528, 30 y/o, 3 children.

James, James Anderson. 25 Aug 1862, conscript, enl. Co. M 1st Palmetto Sharpshooters; des. 25 Aug; 1 Apr 1863, Pickens, SC; pvt.; 20 Aug '63 Columbia, SC; 3 and 12 Nov '63 Atlanta, GA; Nov/Dec '63 pres.; 18 Jan '64 Dalton, GA; 21 Mar '64 clothing; Apr 27 '64 Dalton, GA; 7 May '64 clothing; May '64 des.; 20 Jun '64 Atlanta, GA; Nov. '64 home; 7 Mar '65 abs., resident of Grove Station near Greenville, SC; 1860 Anderson, SC, Regt. 42, #1242, 22 y/o; b. 16 May 1839, d. 2 Nov 1903, bur. Shiloh Cem., Monroe Co., AR.

James, John. 13 Oct 1863 des., Atlanta, GA (letter of his brother, James Anderson); 1860 Anderson, SC, Regt. 42, #1242, 25 y/o; b. Jan 1833; 1910 Anderson, SC, Williamston, Distr. 70, #84, 67 y/o; d. before 1920.

Johnson, Anthony V. 15 Aug 1863, Elbert Co., GA; pvt.; Nov/Dec '63 hosp.; b. 26 Sep 1845, d. 27 Sep 1924, bur. Bethlehem Ch., Elbert, GA.

Jones, John T. 26 Sep 1862 enl. Co. E 31st NC INF, 18 y/o, Wake Co., NC; pvt.; 15 Apr '63 transf. to FAC; Mar/Apr to May/Jun '63 pres.; 15 Jul '63 captd. Jackson, MS, POW Camp Morton, IN to Aug '63, enl. 7th IN CAV, US; NFR.

Jordan, James 20 Feb 1862, New Orleans, LA; pvt.; Apr/Jul '62 to May/Jun '63 pres.; 25 Nov '63 MIA; 26 Nov '63 captd. Ringgold, GA, POW Rock Island, IL to 21 Apr '64, 44 y/o; NFR.

Kay, George Washington 27 Jun 1862, Pickens, SC; pvt.; 15 Jul '62 des.; Jan/Feb to Nov/Dec '63 pres.; Apr and Sep '64 clothing; 12 Apr '65 captd. Salisbury, NC, POW Camp Chase, IL to 13 Jun '65, resident of Anderson Distr., SC; 1870 Pickens, SC, Garvin, #7, 34 y/o; b. 16 Mar 1833, d. 18 Jun 1903, bur. East View Cem., Oconee Co., SC.

Kay, Henry Trussell 23 Feb 1863, Franklin, GA; pvt.; Mar/Apr to Nov/Dec '63 pres.; 4 Jun '64 slight ear wnd., Dallas, GA; Aug '64 clothing; 1900 GA pension; 1860 Franklin, GA, #888, 29 y/o; b. Jan 1831, d. 17 Jan 1915 (pension data), bur. Aycock Cem., Erath Co., TX.

Keefe, Wellington XL Co. L, 8th SC INF to Sep 1863; enl. FAC Oct 1863, Marion, SC; pvt.; 25 Nov '63 MIA; 27 Nov '63 captd. Ringgold, GA, POW Rock Island, IL to 6 Oct '64, enl. Co. F, 2nd US INF for frontier service, fireman, b. 1842, mustered out 7 Nov '65; NFR.

Kennedy, William M. 8 May 1862, GA; pvt.; Apr/Jul '62 pres.; Jul/Aug '62 Summerville hosp., SC; 14 Sep '62 des.; NFR.

* **Kennemore, George** 7 Apr 1863, Pickens, SC; pvt.; Mar/Apr '63 abs.; 29 Jun '63 d. Jackson Hosp., MS; 8 Sep '64 father, Lott, filed claim for deceased soldier; 1860 Pickens, SC, Regt. 5, #1177, 26 y/o. Brother of Jacob.

Kennemore, Jacob 14 Apr 1861 Co. G, 4th SC Volunteers, Pendleton, SC; 24 Oct '61 dc'd., disability, Richmond hosp.; enl. FAC 7 Apr '63, Pickens, SC; pvt.; Mar/Apr '63 not yet reported for duty; May/Jun '63 pres.; 1850 Pickens, SC, East, #300, 16 y/o; 1860 White, GA, #215, 30 y/o; 1900 Pickens, SC, Central, Distr. 117, #124, 73 y/o; NFR. Brother of George

* **Kloster, Charles** 20 Feb 1862, New Orleans, LA; pvt.; Apr/Jul to Jul/Aug '62 pres.; 16 Sep '62 jail, des.; Nov/Dec '62 to May/Jun '63 pres.; 26 Oct '63 d. Five Mile, Pickens, SC, on furlough; NFR.

Ladd, Pleasant Easley 1 Jul 1862, Pickens, SC; pvt.; Jul/Aug '62 to Nov/Dec '63 pres.; Mar, Jun and Aug '64 clothing; 21 May '65 ended service; teamster; 1860 Pickens, SC, Regt. 5, #103, 15 y/o; b. 20 Jan 1845, d. 11 Feb, 1925, bur. Fuller Cem., Henegar, Dekalb Co., AL.

Lafoy, James Harrison 18 Mar 1863, Anderson, SC; pvt.; Mar/Apr to May/Jun '63 pres.; 1 Sep '63 hosp.; Mar, Apr, May, Jun and Aug '64 clothing; 1860 Anderson, SC, Regt. 4, #1037, 14 y/o; 1900 Spartanburg Co., SC, Reidville, Distr. 96, #177, 53 y/o; died April, 1911, bur. Springwood Cemetery, Greenville, SC; VA headstone installed by author; widow's SC pension 1912.

Lawson, Ephraim G. 31 Oct 1861 Pickens, SC; pvt., Co. K, 2nd SC Rifles; head wnd. and ankle GSW; 18 Oct '63 transf. FAC; 25 Nov '63 MIA; Jul '64 hosp., clothing; 12 Apr '65 captd. Salisbury, NC, POW Camp Chase, OH to 13 Jun '65; 1860 Pickens, SC, Regt.

5, #1062, 18 y/o; d. 14 Feb 1916, bur. Oak Grove Bapt. Ch., Tift Co., GA.

Layton, Robert 17 Mar 1863, Darlington, SC; pvt.; Mar/Apr to Nov/Dec '63 pres.; Mar, Jun, Jul and Aug '64 clothing; 1860 Darlington, SC, #584, 13 y/o; d. 20 Jul 1920, bur. Liberty Hill Baptist, Hartsville, SC.

League, William P. 1 Jul 1862, Pickens, SC; pvt.; Apr/Jul '62 to Jul/Aug '62 des.; Jan/Feb to Nov/Dec '63 pres.; Jun '64 clothing; NFR.

* **Leathers, William H., Jr.** 19 Mar 1863, Pickens, SC; pvt.; Mar/Apr to May/Jun '63 pres.; 1 Dec '63 d. hosp. Kingston, GA; father, W.H. Sr. made claim; 1860 Pickens, SC, Regt. 2, #1298, 13 y/o.

Lecluse, Leon 22 Feb 1862, New Orleans, LA; pvt.; Apr/Jul to Jul/Aug '62 hosp. Columbia, SC; Sep/Oct '62 to Jan/Feb '63 pres.; Mar/Apr '63 sick in camp; May/Jun '63 sick, Summerville, SC hosp.; Nov/Dec '63 hosp., diarrhea; Apr '64 clothing; 21 Jul to 9 Aug '64, hosp. Mobile, AL, dysentery; 28 Dec '64 Lumpkin Hosp., Cuthbert, GA; 26 Apr '65 captd. Greensboro, NC, took oath 15 Jun, Mobile, AL, 49 y/o; NFR.

Lecuntaguy, Victor 28 Aug 1862, Columbia, SC; pvt.; Jul/Aug '62 to Jan/Feb '63 pres.; Mar/Apr '63 pres., cpl. since 1 Mar; CM Apr/May '63; May/Jun '63 pres.; Jul '63 captd. Jackson, MS, POW Camp Morton, IN to 17 Aug '63, 25 y/o, enl. 1st Regt, Baty. M, MI Lt Arty., US, mustered out 19 Mar '64; 1878 Galveston, TX city directory "V. E. Lacuntiguy;" NFR.

* **Lee, Noah** 30 Mar 1863, Pickens, SC; pvt.; Mar/Apr to May/Jun '63 pres.; Nov/Dec '63 hosp. since Apr Point Clear, Baldwin Co., AL; 1850 Pickens, SC, West, #655, 4 y/o; d. Marietta, GA, date not known, bur. Marietta Confederate Cemetery.

Mann, Samuel P. 9 Apr 1863, Pickens, SC; pvt.; Mar/Apr '63 abs.; May/Jun to Nov/Dec '63 pres.; May and Aug '64 clothing; 1860 Pickens, SC, Regt. 5, #907, 14 y/o; NFR.

Marie, Jean 12 Mar 1862, New Orleans, LA; pvt.; Apr/Jul '62 pres.; 16 Jul AWOL, Jul/Aug '62 pres.; Sep/Oct '62 jail, des.; Nov/Dec '62 jail 20 Nov, pres.; Jan/Feb '63 pres.; Mar/Apr '63 jail since Apr 20, des., CM; May/Jun '63 jail; Nov/Dec '63 pres.; Mar, May, Jun, Jul and Aug '64 clothing; NFR.

* **Martin, Benjamin** 23 Jun 1863, Columbia, SC; pvt.; Nov/Dec '63 hosp.; Mar and Apr '64 clothing; 12 Apr '65 captd. Salisbury, NC, POW Camp Chase, OH; 2 Jul '65 d. Camp Chase, bur. there in Confederate cem.; 1860 Anderson, SC, #979, 20 y/o.

Mason, John 15 Mar 1862, New Orleans, LA; pvt.; 4 Jun '62 dc'd., varicose veins, 45 y/o b Ireland; NFR.

Masters, John Calhoun enl. unjustly in Aiken Partisan Rangers, Maguire's, 6th SC CAV, returned to FAC, 1 Jul 1862, Pickens, SC; pvt.; Sep/Oct '62 pres.; Nov/Dec sick furlough 1 Dec; Jan/Feb to Nov/Dec '63 pres.; Aug '64 clothing; May '65 at Salisbury, NC; 1880 Greenville, SC, Cleveland, 95, #59, 34 y/o; b. 12 May 1844, d. 14 Mar 1933, bur. Saluda Hill Baptist Ch., Pickens Co., SC.

McCauley, Daniel 18 Feb 1862, New Orleans, LA; pvt.; Apr/Jul '62 disability discharge, 38 y/o Irish shoemaker, inguinal hernias; NFR.

McCormick, William 22 Aug 1863, Columbia, SC; pvt.; Nov/Dec '63 pres.; Jun and Aug '64 clothing; blacksmith; NFR.

McCullough, Hiram A. 2 May 1862, Charleston, SC; pvt.; Apr/Jul '62 pres.; Jul/Aug '62 hosp. Summerville, SC; Sep/Oct '62 to Nov/Dec '63 pres.; 1 Jan '64 at Dalton, GA; Mar, Apr, Jul and Aug '64 clothing; 1850 Harris Co., GA, #997, 10 y/o; b. GA 12 Jul 1840, d. 23 May 1915, bur. Rosemere Cem., Opelika, AL.

McJunkin, John D. 10 Oct 1861 4th SC Bttn; 1 Jul '62, Pickens, SC; pvt.; Jan/Feb '63 pres.; 29 Mar to Dec '63 AWOL; 1st SC State Troops 1 Aug '63; 2nd Rifles 20 Feb '64; 1860 Pickens, SC, Regt. 5, #188, 36 y/o, b. Buncombe Co., NC; d. 11 Jun '64 Richmond, VA hosp.

* **McMartin, Jasper** 6 Mar 1862, New Orleans, LA; pvt.; Apr/Jul to Jul/Aug '62 hosp. Columbia, SC; Sep/Oct '62 to Mar/Apr '63 pres.; d. 14 May '63 Jackson, MS.

McNalty, John R. 20 Feb 1861, Memphis, TN enl. 1st SC INF (Butler's), 30 y/o; made sgt., reduced to pvt.; 1 Jan '63 transf. to FAC in exchange for John Scott; Jan/Feb to Mar/Apr '63 abs., recruiting, sgt.; May/Jun '63 pres.; Nov/Dec '63 Atlanta jail, CM, reduced to pvt.;

Feb '64 clothing; captd. 5 Jul '64 Marietta, GA, POW Camp Douglas, IL to 13 Mar '65; NFR.

Meredith, Dudley 23 Apr 1863; pvt.; Mar/Apr to May/Jun '63 pres.; Nov/Dec '63 abs., hosp. 11 Jul; Jun, Jul and Nov '64 (hosp. Columbia, SC) clothing; 1860 Pickens, SC, Regt. 5, #1070, 35 y/o, 3 children; 1901 pension list, Central, Pickens, SC; NFR.

Mitchell, Robert L. 10 Feb 1862, Huntsville, AL; pvt.; Apr/Jul to Jul/Aug '62 pres.; Sep/Oct '62 to Jan/Feb '63 pres., cpl.; Mar/Apr to Nov/Dec '63 pres., prom. sgt. 1 Mar; Apr, May, Jun and Aug '64 clothing; 1860 Chambers Co., Lafayette, AL, #89, 22 y/o law student; b. 13 Aug 1837, d. 4 Jul 1872, bur. Tuskegee City Cem., Macon Co., AL.

* **Moran, John** 18 Feb 1862, New Orleans, LA; pvt.; 28 Apr '62 d. Roper Hosp., Charleston, SC.

Morgan, Allen R. 4 Mar 1862, Pickens, SC; pvt.; Apr/Jul to Sep/Oct '62 pres.; Nov/Dec '62 sick furlough 19 Dec; Jan/Feb to May/Jun '63 pres., cpl.; 25 Nov wnd. Missionary Ridge, TN, furloughed; farrier; 1870 Pickens, SC, Garvin, #245, 39 y/o, 4 children in 1862; 1900 Anderson Co., SC, Williamston 69 #220, 67 y/o; NFR.

* **Morgan, Charles C.** 2 Apr 1863, Pickens, SC; pvt.; Mar/Apr '63 abs.; May/Jun '63 pres.; d. hosp., Kingston, GA, widow, Sarah 1 Dec '64 claim for deceased soldier.

Morgan, Lewis W. 2 Apr 1863, Pickens, SC; pvt.; Mar/Apr '63 abs.; May/Jun to Nov/Dec '63 pres.; Apr, May and Jun '64 clothing; 19 Dec '64 captd. Franklin, TN, POW Nashville, TN to May; 16 Jun '65 took oath at Louisville, KY prison, resident Walhalla, SC; 11 to 24 Aug '65 USA hosp., Augusta, GA, fever; 1860 Pickens, SC, Regt. 2, Oconee Station, #373, 32 y/o; b. 10 Sept 1828, d. 28 July 1892, bur. Fairview Comm. Ch., Oconee, SC.

Morgan, R.P. 1850 Pickens, SC, Western Div., #1120, 8 y/o; 1900 Pickens, SC, Hurricane, #382, b. May 1840, 60 y/o; 1901 pension, Alexander, Pickens Co., SC, 60 y/o, Beauregard's Arty.; NFR.

Murphy, Dennis 10 Nov 1861, Summerville, SC enl. Co. B Palmetto Btn. Arty. (Watie's); pvt.; 5 Apr '62 des.; Sep/Oct '62 to Jan/Feb '63 Charleston jail; 11 Mar '63 transf. FAC; Mar/Apr '63, pres., cpl.; May/Jun to Nov/Dec '63 abs., 10 May 1st MS hosp., Jackson, then 1 Jun '63 to Gen. hosp.; NFR.

Murphy, Michael 3 Mar 1862, New Orleans, LA; pvt.; Apr/Jul '62 pres.; Jul/Aug '62 hosp. Summerville, SC; Sep/Oct '62 to May/Jun '63 pres.; 29 Aug '63 AWOL; NFR.

Murray, Warren W. 3 Apr 1863 enl. Co. E, 62nd NC INF, Brevard, NC; des. 10 May at Greenville; 6 Jul 1863 enl. FAC, Walhalla, SC; pvt.; Nov/Dec '63 pres.; Jan/Feb '64 Blackie Hosp., Madison, GA; Apr, Jun and Aug '64 clothing; 12 Apr '65 captd. Salisbury, NC, POW Camp Chase, OH to 13 Jun, 19 y/o resident of Transylvania, NC; 1860 Pickens, SC, Regt. 2, #23, 19 y/o; d. 28 May 1926, bur. Gashes Creek Cem., Asheville, NC; VA headstone installed by author.

Myers, Henry 2 Apr 1863, Pickens, SC; pvt.; Mar/Apr to Nov/Dec '63 abs., not yet reported to company; tentative 1860 Pickens, SC, Regt. 2, #1491, 37 y/o; 1870 Oconee Co., SC, Center, #1, 47 y/o; NFR.

Nix, James 29 Apr 1863, Pickens, SC; pvt.; Mar/Apr '63 not yet reported; May/Jun '63 pres.; Nov/Dec '63 hosp. since 1 Sept; either 1860 Pickens, SC, Regt. 2, #230, 28 y/o, 2 children, or 1860 Pickens, SC, Regt. 5, #123, 25 y/o; NFR.

O'brien, John 12 Mar 1862, New Orleans, LA; pvt.; Apr/Jul '62 to May/Jun '63 pres.; Nov/Dec '63 abs., MIA 25 Nov; 26 Nov captd. Ringgold, GA, POW Rock Island, IL to 21 Apr '65, 25 y/o; NFR.

O'Kieff, David 15 Mar 1862, New Orleans, LA; pvt.; 4 Jun '62 disability dc., loss of an eye and cardiac hypertrophy, 42 y/o Irish shoemaker; NFR.

* **Oliver, J. Walter** 29 Oct 1862, Sullivan's Island, SC; pvt.; Sep/Oct '62 to Mar/Apr '63 pres.; 25 May '63 d. typhoid pneumonia, 1st MS Hosp., Jackson, MS.

Partains 1 Jul 1862, Anderson Distr.; pvt.; Jul/Aug '62 pres.; 8 Sep '62 des.; NFR.

Passell, Charles 24 Feb 1862; New Orleans, LA; pvt.; Apr/Jul '62 pres.; Jul/Aug '62 hosp. Summerville, SC; Sep/Oct '62 det. Naval service; NFR.

Patterson, John 3 Feb 1862, Charleston, SC; pvt.; Apr/Jul '62 to May/Jun '63 pres., extra duty mending harness; 25 Nov '63 MIA; 26 Nov '63 captd. Ringgold, GA, POW Rock Island, Il to 22 Apr '64, 42 y/o, 18 Mar '64 asked to take oath; rejected in bid to join US Navy; Rebel POW of this name did serve on USS Kanawha Feb to Jun '65; NFR.

Phillips, Nathaniel 16 Sep 1863, Pickens, SC; pvt.; Nov/Dec '63 pres.; Mar and Apr '64 clothing; 1 Sep '64 des.; 1860 Pickens, SC, Regt. 2, #966, 15 y/o; b. 21 May 1845; 1915 SC pension (suffered side wound); d. 17 Jan 1920, bur. Phillips Family Cem., Long Creek, Oconee Co., SC.

Porter, Carter Clayton 6 Apr 1863, Pickens, SC; pvt.; Mar/Apr '63 not yet reported; May/Jun to Nov/Dec '63 pres.; Mar '64 clothing; Sep '64 pd.; 12 Apr '65 captd. Salisbury, NC, POW Camp Chase, OH to 13 Jun '65, 38 y/o; 1850 Pickens, SC, Eastern Div., #8, 21 y/o; b. 24 Dec 1827, d. 2 Jun 1920, bur. Jones Memorial Cem., Clayton, GA. Brother of William C.

Porter, Perry Warren 15 Feb 1863 enl. 65th GA INF, Loudsville, GA; pres. to Aug '64; 1 Nov '64 transf. to FAC; 16 Dec '64 captd. Nashville, TN, POW Camp Douglas, IL to 22 Jun '65; 1860 Lumpkin, GA, Crumbys, #707, 14 y/o; b. 3 Aug 1845, d. 5 Dec 1931, bur. Cherry Grove Cem., Palestine, Cleburne Co., AL.

Porter, William C. 25 Dec 1862, White Co., GA; pvt.; Jan/Feb '62 to Nov/Dec '63 pres.; Apr and May '64 clothing; 12 Apr '65 captd. Salisbury, N.C., POW Camp Chase, OH to 13 Jun, 42 y/o resident of White Co., GA; 1870 White Co., GA, Mossy Creek, #109, 47 y/o; b. 23 May 1823, d. 21 Apr 1911, bur. Alta Vista Cem., Hall, GA. Brother of Carter Clayton.

* **Prince, John F.** 7 Apr 1863, Pickens, SC; pvt.; Mar/Apr '63 not yet reported; May/Jun '63 pres.; 25 May MIA; 26 Nov '63 captd. Ringgold, GA, POW Rock Island, IL, d. there 9 Jan '64, bur. #169 Rock Island Confed. Cem.; 8 Sep '64 widow, Susan, made claim for deceased soldier; 1860 Pickens, SC, Regt. 5, #1174, 30 y/o, 4 children.

Prince, William Patrick 8 Aug 1863, Abbeville, SC; pvt.; Nov/Dec '63 pres.; Mar and May '64 clothing; 15 May '64 slight wnd., side,

Resaca, GA; 1860 Abbeville, SC, #2014, 31 y/o, 3 children; d. 16 Aug 1906, bur. Good Hope Presb. Ch., Iva, Anderson, SC.

Quinn, John F. 23 Jan 1862, Charleston, SC; pvt.; Apr/Jul '62 abs., recruiting; Jul/Aug '62 pres.; Sep/Oct '62 sgt., abs., recruiting; Nov/Dec '62 to Jan/Feb '63 pres.; Mar/Apr '63 recruiting; May/Jun '63 pres.; Nov/Dec '63 furlough 21 Dec; Apr, May, Jun, Jul and Aug '64 clothing; Dec '64 captd. Nashville, TN, POW Camp Douglas, IL to 12 May '65, claimed loyalty and took oath, Irish, resident Charleston, SC; 1850 St. Michael Parish, Charleston, SC #634, 11 y/o; d. 2 Apr 1867, orchitis, 27 y/o, bur. Saint Lawrence Cem., Charleston, SC; VA headstone provided by author.

Quinn, William 4 Feb 1862, Charleston, SC; pvt.; Apr/Jul to Jul/Aug '62 pres., sgt. 1 Jun; Sep/Oct '62 abs., pvt.; 14 Sep des.; Nov/Dec '62 jail since 2 Nov; Jan/Feb '63 pres.; Mar/Apr '63 abs., 22 Apr 15 days jail in solitary per CM, released 6 May; May/Jun to Nov/Dec '63 pres., returned from jail 6 May; 17 May '64 missing at Calhoun, GA; NFR.

Rains, G.W. 20 Feb 1862, New Orleans, LA; pvt.; Apr/Jul '63 hosp., Columbia, SC; Jul/Aug '62 pres.; Sep/Oct '62 abs., 15 Sep des.; Nov/Dec '62 to Jan/Feb '63 pres.; Mar/Apr '63 abs., recruiting; May/Jun '63 pres.; NFR.

Ramey, Albert Mar, Apr and Aug 1864, clothing; 1860 Rabun Co., GA, Chechero #416, 12 y/o; b. May 1848; 1914 SC pension; d. 4 May 1923, bur. Holly Springs Cem., Westminster, Oconee Co., SC.

Reaves, Joseph 6 Apr 1863, Pickens, SC; pvt.; Mar/Apr to Nov/Dec '63 pres.; 1860 Pickens, SC, Regt. 5, #29, 21 y/o; b. 1 Jan 1836, Spartanburg, SC; 1901 SC pension; d. 25 Jun 1916, bur. Cedar Rock Bapt. Ch., Easley, Pickens, SC.

* **Rhoden, H.** 8 Aug 1863, Columbia, SC; pvt.; Nov/Dec '63 pres.; d. in service, date unknown, receipt for personal effects $31.50 filed in 1864; NFR.

Rhoden, W. 15 Jul 1863, Columbia, SC; pvt.; Nov/Dec '63 recruiting, abs.; NFR.

Rigdon, Benson (Benjamin) H. 8 Apr 1863, Pickens, SC; pvt.; Mar/Apr '63 not yet reported; May/Jun '63 pres.; Nov/Dec '63 abs.,

AWOL 14 Jul; 18 Jul '63 captd. Brendan, MS, POW Camp Morton, IN to 16 May '65, claimed to have been conscripted; 1860 Pickens, SC, Regt. 5, #109, 23 y/o, 2 children; 1900 IL, Perry, Pinckneyville, distr. 55, #288, 65 y/o, b. June, 1834; NFR.

Robier, Phillip 21 Feb 1862, New Orleans, LA; pvt.; Apr/Jul '62 pres.; Jul/Aug '62 hosp. Summerville, SC; Sep/Oct '62 to May/Jun '63 pres.; Nov/Dec '63 MIA 25 Nov; 26 Nov '63 captd. Ringgold, GA, POW Rock Island, IL to 25 Jan '64, enl. US Navy, transf. to Naval Rendezvous, Camp Douglas, IL, served on USS Canonicus 23 Apr '64; NFR.

Russell, William A. 18 Mar 1862, Pendleton, Pickens, SC; pvt.; Apr/Jul to Jul/Aug '62 pres.; Sep/Oct '62 abs., det. serv.; Nov/Dec '62 pres.; Jan/Feb '63 abs., det. serv. 27 Feb; Mar/Apr to Nov/Dec '63 pres.; Mar '64 clothing; NFR.

Sanders, James 10 Mar 1863, Marlboro, SC; pvt.; Mar/Apr '63 pres.; May/Jun to Nov/Dec '63 abs., AWOL since Apr; POW Camp Chase, OH to 13 Jun '65; NFR.

Sargent, Henry Warren 2 Jun 1861 enl. 4th SC INF, Pendleton, Pickens, SC; pvt., 23 y/o; 2 Nov '61 dc'd. for disability GSW head 29 Aug, Manassas, VA; 4 Dec '61 enl. 1st SC CAV, Pickens, SC; 28 Dec '62 dc'd. for disability blind one eye, partial blindness the other; 8 Apr '63 enl. FAC, Pickens, SC; Mar/Apr to May/Jun '63 not yet reported; Nov/Dec '63 pres.; May '64 clothing; 1860 Pickens, SC, Regt. 5, #1044, 22 y/o; b. 1838, d. 1896, bur. Piney Grove Bapt. Cem., Dekalb Co., AL. All Sargents are brothers.

Sargent, John Bailes 13 Jan 1862 enl. 2nd SC Rifles, Pickens, SC; pvt.; transf. after Aug '63 to FAC; Nov/Dec '63 pres.; Jun '64 clothing; 1860 Pickens, SC, Regt. 5, #1044, 16 y/o; b. 28 Jan 1844, d. 19 Dec 1912, bur. Flat Rock Bapt. Ch., Liberty, Pickens Co., SC. All Sargents are brothers.

Sargent, Jesse H. 9 Mar 1863, Pickens, SC; pvt.; Jan/Feb to Nov/Dec '63 pres.; Jun and Aug '64 clothing; 12 Apr '65 captd. Salisbury, NC, POW Camp Chase, OH to 13 Jun; 1860 Pickens, SC, Regt. 5, #895, 28 y/o; b. 1829, d. 21 Dec 1880, bur. Beaverdam Bapt. Ch., Williamston, Anderson Co., SC. All Sargents are brothers.

Sargent, William H. 4 Mar 1862, Pickens, SC; cpl.; Apr/Jul to Sep/Oct '62 pres., cpl. 1 Jun; Nov/Dec '62 to Jan/Feb '63 pres., sgt.; Mar/Apr '63 abs., recruiting; May/Jun to Nov/Dec '63 pres.; Jun and Aug '64 clothing; 12 Apr '65 captd. Salisbury, NC, POW Camp Chase, OH to 13 Jun; 1860 Pickens, SC, Regt. 5, #1056, 32 y/o; b. 1828, d. 1872 Pickens, SC. All Sargents are brothers.

Saxton, A.C. 15 Jul 1863, Columbia, SC; pvt.; Nov/Dec '63 pres.; Mar/Jul '64 clothing; NFR.

Saxton, John D. 15 Aug 1863, Abbeville, SC; pvt.; Nov/Dec '63 pres.; Apr, Jun and Aug '64 clothing; NFR.

Scott, J. John 1 Nov 1862, Pickens, SC; pvt.; Sep/Oct to Nov/Dec '62 pres.; 1 Jan '63 transf. 1st SC INF (Butler's) in exchange for John R. McNalty; Feb '64 transf. 22nd SC INF; 30 Jul '64 "casualty of war" Petersburg, VA; 1860 Pickens, SC, Regt. 2, #225, 30 y/o day laborer.

Scott, Jackson 16 Sep 1863, Columbia, SC; pvt.; Nov/Dec '63 pres.; Mar '64 clothing; NFR.

Seawright, Enoch Wilson 10 Feb 1863, Abbeville, SC; pvt.; Jan/Feb to Nov/Dec '63 pres.; 12 Apr '65 captd. Salisbury, NC, POW Camp Chase, OH to 13 Jun '65; 1860 Abbeville, SC #2123, 15 y/o; b. 11 Mar 1845, d. 21 Jul 1916, bur. Greenville Presb. Ch. Cem., Shoals Junction, Greenwood Co., SC.

Sears, Thomas Glenn 25 Dec 1862, White Co., GA; pvt.; Jan/Feb to Nov/Dec '63 pres.; 12 Apr '65 captd. Salisbury, NC, POW Camp Chase, OH to 13 Jun '65, 41 y/o; 1860 White Co., GA #233, 35 y/o, 5 children; b. Oct 1824, d. 6 Oct 1901, bur. Wesley Chapel Cem., Union, Fulton Co., AR.

Shettlesworth, Ira A. 12 Jan 1863, Pickens, SC; pvt.; Jan/Feb to May/Jun '63 pres.; Nov/Dec '63 abs., hosp. 11 Jun; Mar '64 clothing; 4 May '65 captd., Citronelle, AL, POW Meridian, MS to 12 May; 1880 Franklin Co., GA, Big Smiths, 30, #3, 43 y/o; 1883 GA Property Tax Records, Franklin Co.; NFR. Probable brother of Jackson Burnett.

Shettlesworth, Jackson Burnett 12 Jan 1863, Pickens, SC; pvt.; Jan/Feb to Mar/Apr '63 pres.; May/Jun '63 abs., hosp. 11 Jun; 13 Jul '63 captd., Yazoo City, MS, paroled; Nov/Dec '63 pres.; Mar, Jun and Jul '64 clothing; 12 Apr '65 captd., Salisbury, NC, POW Camp

Chase, OH to 13 Jun '65, 20 y/o; 1860 Pickens, SC, Regt. 2 #806, 16 y/o; 1890 Whitfield, GA tax roll; NFR. Probable brother of Ira A.

Simons, 'Iön 25 Mar 1862, Grahamville, SC enl. 4th SC CAV, 21 y/o; pvt.; promoted cpl. by Jun '63; 1 Jul '63 appointed 2nd lt. FAC; May/Jun '63 abs., recruiting in Pickens, SC; 8 Dec '64 rank confirmed; 1 Jan '65 rank accepted; 1860 Charleston, SC, Ward 2, #136, 20 y/o clerk; b. 3 Sep 1839, d. 26 Jan 1899, bur. Magnolia Cem., Charleston, SC.

Sizemore, William 25 Mar 1862, Pickens, SC; pvt.; Apr/Jul to Jul/Aug '62 pres.; Sep/Oct '62 abs., des. 12 Sep; Nov/Dec '62 pres.; CM Jan '63, sentence remitted Feb; Mar/Apr to May/Jun '63 pres.; 25 Nov '63 MIA; 27 Nov '63 captd., Ringgold, GA, POW Rock Island, IL to 6 Oct '64, enl. Co. D, 2nd USVI, pvt., frontier service, dc'd 1 Nov '65; 1870 Pickens, SC, Pickens, #174, 48 y/o; 1890 Vet Schedule, Pickens, SC; b. Jan 1827, d. 15 Jan 1905; bur. Cold Spring Baptist Ch., Pickens, SC; church declined to allow VA headstone.

Smart, John T. 6 Apr 1863, Pickens, SC; pvt.; Apr/May '63 Soldiers Rest Hosp., Mobile, AL; May/Jun '63 pres.; 27 Jul '63 AWOL; Aug '63 pd.; 1850 Laurens, SC, #2027, 5 y/o; d. 22 Dec 1910, bur. Elmwood Cem., Columbia, SC, square 41; VA headstone provided by reenactors of Ferguson's SC Artillery Company.

Smith, B.H. 12 Aug 1863, Abbeville, SC; pvt.; Nov/Dec '63 pres.; Jun '64 clothing; NFR.

Smith, E.N. 26 Jun 1862, Abbeville, SC; pvt.; Mar/Apr '63 pres., AWOL from enl. to 26 Apr '63; May/Jun '63 pres.; Nov/Dec '63 hosp. since 26 Nov; NFR.

* **Smith, Freeman** 24 Feb 1863, Anderson, SC; pvt.; Jan/Feb to May/Jun '63 pres.; Nov/Dec '63 abs., hosp. since 21 Oct; 20 Nov '63 pd., Rome, GA; d. Abbeville, SC before 2 Apr '64, when his widow, Susan E,, made claim.

Smith, John 28 Jun 1862, Anderson, SC; pvt.; Jul/Aug '62 hosp.; Sep/Oct '62 to Nov/Dec '63 pres.; 12 Apr '65 captd., Salisbury, NC, POW Camp Chase, OH to 13 June, 26 y/o, Clark Co., GA; NFR.

Southerland, J.B. 24 Jun 1864 clothing; NFR.

Spears, Asbury F. Apr 1862 Hart Co., GA militia; enl. FAC 2 Mar 1863, Hart Co., GA; pvt.; Mar/Apr to Nov/Dec '63 pres.; Aug '64 teamster; Sept '64 clothing; 1860 Anderson, SC, Regt. 42, #578, 3 children; b. 13 Jul 1833, d. 24 Oct 1904, bur. Old Canon Bapt. Ch., Hart Co., GA; VA headstone installed by author.

Spears, Samuel E. 16 Mar 1863, Darlington, SC; pvt.; Mar/Apr to May/Jun '63 pres.; 1860 Darlington, SC, #636, 23 y/o; 1891 pension application; 1912 pension, wife, Emily; NFR.

* **Spencer, James M.** 8 May 1862, Pickens, SC; pvt.; Jul/Aug '62 Summerville, SC hosp.; Sep/Oct '62 pres.; 22 Nov '62 d. in camp, Mt. Pleasant, SC; 11 Mar '64 widow, Amanda, claim for deceased soldier; 1860 Pickens, SC, Regt. 2, #1281, 30 y/o, 1 child; bur. Hopewell U. Meth. Ch. Cem., Westminster, Oconee Co., SC. Brother of Joshua.

* **Spencer, Joshua** 8 May 1862, Pickens, SC; pvt.; Apr/Jul '62 pres.; Jul/Aug '62 hosp. Summerville, SC; Sep/Oct '62 abs., sick, Pickens, since 31 Aug; 10 Jan '63 d. Sullivan's Isl., SC; 2 Feb '63 requisition for his coffin, 6'2" long; 11 Mar '64 mother, Soleana, claim for deceased soldier; 1860 Pickens, SC, Regt. 2, #1279, 18 y/o; bur. Hopewell U. Meth. Ch. Cem., Westminster, Oconee Co., SC. Brother of James M.

St John, William H. 22 Jun 1863, Columbia, SC; pvt.; Nov/Dec '63 pres.; Apr, Jun and Sep '64 clothing; NFR.

* **Stancil, Bailey.** 21 Feb 1863, Pickens, SC; pvt.; Jan/Feb to May/Jun '63 pres.; Nov/Dec '63 abs., French's Div. Hosp., Lauerdale, MS since 15 Jul; 20 Oct '64 Gen. Hosp., Selma, AL, ulcers and chronic diarrhea; d. 12 May (dates inconsistent); 1860 Pickens, SC, Regt. 5, #134, 18 y/o. Brother of Eli and James Ervin.

Stancil (Stancill), Eli 1 Jul 1862, Pickens, SC; pvt.; Jul/Aug '62 pres.; Sep/Oct '62 sick Summerville, SC hosp. since 6 Oct; Nov/Dec '62 30 day sick furlough since 22 Dec; Jan/Feb to Nov/Dec '63 pres.; May, Jun and Jul '64 clothing; 12 Apr '65 captd. Salisbury, NC, POW Camp Chase, OH to 13 Jun '65; 1897 GA pension application, GSW and deaf left ear; 1860 Pickens, SC, Regt. 5, #155, 23 y/o; b. 27 May 1835, d. 15 Jun 1899, bur. Walnut Fork Cem., Braselton, Jackson Co., GA. Brother of Bailey and James.

* **Stancil, James Ervin** 21 Feb 1863, Pickens, SC; pvt.; Jan/Feb to May/Jun '63 pres.; 25 Nov '63 MIA; 27 Nov '63 captd., Ringgold,

GA, POW Rock Island, IL, 19 Dec '63 d. Rock Island, pneumonia, bur. south of barracks #26 (Joseph A. Stancel); 1860 Pickens, SC, Regt. 5, #134, 14 y/o. Brother of Baily and Eli.

Stansell, Thomas J. 2 Jul 1862, Pickens, SC; pvt.; Jul/Aug '62 sick, Summerville, SC hosp.; Sep/Oct '62 sick, Summerville hosp. since 5 Sep; Nov/Dec '62 to Nov/Dec '63 pres.; Apr '64 clothing; 23 Dec '64 des.; 7 Mar '65 abs.; 6 Oct 1929 pension states served to May '65, dc'd. at Salisbury, NC; 1850 Pickens, SC, East, #508, 6 y/o; b. 1 Jun 1844, d. 16 Jun 1924, bur. Mt. Carmel Bapt. Ch. Cem., Easley, Pickens Co., SC.

Stark, Louis Raoul 28 Apr 1862, 2nd Military District, SC, enl. in Marion Arty.; pvt.; 6 Jun '62 pres.; 17 Jun '62 appointed 2nd lt, FAC; Apr/Jul to Jul/Aug '62 pres; Sep/Oct '62 enrolling conscripts, Laurens, SC 1 Oct; Nov/Dec '62 witness, CM, Charleston, SC; Jan/Feb to Mar/Apr '63 det. serv. 27 Feb; May/Jun '63 pres. since 8 May; 4 Jan '64 resigned to be Adj., 10th SC CAV; 1860 Charleston, SC, Ward 5, The Citadel, #79, 18 y/o student; b. 4 Jun 1841, d. 10 Apr 1909, bur. Mt. Holly Cem., Little Rock, Pulaski Co., AR.

Steatland, Charles 3 Mar 1862, New Orleans, LA; pvt.; Apr/Jul '62 to Nov/Dec '63 pres.; Jun, Jul and Aug '64 clothing; 12 Apr '65 captd. Salisbury, NC, POW Camp Chase, OH to 13 Jun, 20 y/o resident of New Orleans; NFR.

* **Stewart, Robert** 6 Apr 1863, Pickens, SC; pvt.; Mar/Apr '63 not yet reported; May/Jun '63 pres.; Nov/Dec '63 abs., hosp. since 1 Oct; 1860 Pickens, SC, Regt. 5, #425, 24 y/o, 2 children; d. Atlanta, GA hosp., widow, Elmira E., claim for deceased soldier 27 Jul '64; possibly d. 7 Jun '64, bur. Oakland Cem., Atlanta, Fulton Co., GA.

Stewart, Whitfield 1 Mar 1862, Newberry, SC, enl. 3rd SC INF, 19 y/o; 29 Jul '62 dc'd., pulmonary consumption, Richmond, VA; 22 Jun 1863, enl. FAC, Columbia, SC; pvt.; Nov/Dec '63 pres.; 14 Jan '64 des.; 4 Oct '64 clothing; b. Aug. 1843, d. 19 Apr 1920, Newberry Co., SC, bur. Zoar UM Ch., Saluda, SC; VA headstone installed by author.

Stone, Beverly Jarett. 13 May 1862, Habersham, GA; pvt.; Apr/Jun '62 pres.; Jul/Aug '62 hosp., Summerville, SC; Sep/Oct '62 pres.; Nov/Dec '62 to Jan/Feb '63 sick furlough 29 Nov; Mar/Apr to

May/Jun '63 AWOL; Nov/Dec '63 hosp. since Dec; 14 Dec '63 and Mar '64 clothing; Jun '64 Ocmulgee Hosp., Macon, GA, anemia; 5 Sep '64 Walker Hosp., Macon, GA, varicose veins; 8 Aug '64 clothing; shoemaker; 1860 Habersham, GA #534, 21 y/o, 2 children; 3 May 1893 GA Confederate pension, leg sores; b. 24 Aug 1838; d. 3 May 1893; NFR.

* **Stone, Joseph** 18 Mar 1862, Pickens, SC; pvt.; Apr/Jul '62 pres.; 9 Aug '62 d. hosp., Summerville, SC; widow, Mahaly Stone; 1850 Pickens, SC, Western Div., #354, 10 y/o.

Strickland, S.E. 20 Feb 1863, Abbeville, SC; pvt.; Jan/Feb '63 pres.; Mar/Apr '63 sick in camp; May/Jun '63 Gen. Hosp. No. 1, 22 May; Nov/Dec '63 hosp.; NFR.

Sullivan, Charles B. 5 Mar 1863, Hart Co., GA; pvt.; Mar/Apr to May/Jun '63 pres.; Nov/Dec '63 AWOL 4 Sep; 16 Dec '64 captd., Nashville, TN, POW Camp Douglas, IL to 16 May '65, took oath, claimed to be a conscript, loyal to US, des.; 1860 Greenville, SC, Sterling Grove, #1588, 33 y/o, 5 children; b. 1824, d. 13 Sep 1891, bur. Townville Bapt. Ch., Anderson, SC.

Summerell, Pleasant Mattison. 12 Apr 1865 captd., Salisbury, NC, POW Camp Chase, OH to 13 Jun '65, 27 y/o, resident of Pickens, SC; 1860 Pickens, SC, Regt. 5, #1285, 25 y/o; b. Aug 1834, d. 14 Apr 1920, influenza, bur. New Hope Ch., Seneca, Oconee, Co., SC.

Taylor, James 22 Feb 1862, New Orleans, LA; pvt.; Apr/Jul '62 pres.; Jul/Aug '63 Summerville, SC hosp.; Sep/Oct '62 to Nov/Dec '63 pres.; 1 Jan '64 St Mary's Hosp., La Grange, GA; May/Jun '64 duty at Atlanta, GA military prison; 24 Jun '64 clothing; 20 Apr '65 captd., Macon, GA; NFR.

Terry, John Bartley pvt.; Jun, 1864, clothing; 12 Apr '65 captd., Salisbury, NC, POW Camp Chase, OH, 19 y/o, resident of Elbert Co., GA; 1880 Anderson, SC, Garvin, #387, 30 y/o; b. 23 Aug 1845, d. 27 Oct. 1918, bur. Tranquil Meth. Ch., Greenwood, SC.

Thomas, J.P. 23 Mar 1863, Anderson, SC; pvt.; Mar/Apr to Nov/Dec '63 pres.; Jun and Aug '64 clothing; NFR.

Turner, Thomas 1 Sep 1863, Edgefield, SC; pvt.; Nov/Dec '63 pres.; 7 May '64 clothing; NFR.

Vocaline, Charles 22 Feb 1862, New Orleans, LA; pvt.; Apr/Jul '62 disability dc., 41 y/o French grocer, organic heart disease; NFR.

Walker, Benjamin F. 25 Feb 1863, Pickens, SC; Mar/Apr to Nov/Dec '63 pres.; Mar and May '64 clothing; 1850 Anderson, SC, East, #693, 8 y/o; b. 17 Jul 1843, d. 27 Aug 1911, bur. Dorchester Cem., Belton, Anderson Co., SC. Brother of Christopher Columbus.

Walker, Christopher Columbus 1 Jul 1862, Pickens, SC; pvt.; Nov/Dec '62 to Nov/Dec '63 pres.; Mar, May, Jun, Jul and Aug '64 clothing; 1850 Anderson, SC, East, #693, 10 y/o; b. 9 Feb 1840, d. 19 Jan 1913, bur. Silver Brook Cem., Anderson Co., SC. Wife's SC pension 1914. Brother of Benjamin F.

* **Ward, Bartlett** 31 Aug 1863, Columbia, SC; pvt.; Nov/Dec '63 pres.; d. in service, $16.60 returned to quartermaster in 1864.

Watson, Charles 12 Apr 1865 captd., Salisbury, NC, POW Camp Chase, OH to 13 Jun '65, 29 y/o resident of Rowan Co., NC; NFR.

Webb, Nelson R. 23 Mar 1863, Marlboro, SC; pvt.; transf. to Co. E, 31st NC 15 Apr '63; b. 1818; NFR.

Weinberg, Andrew 21 Feb 1862, New Orleans, LA; pvt; Apr/Jul '62 pres.; Jul/Aug '62 abs., in Summerville, SC hosp.; Sep/Oct '62 pres.; Nov/Dec '62 pres.; Jan/Feb '63 pres.; Mar/Apr '63 pres.; May/Jun '63 pres.; 25 Nov '63 MIA; 26 Nov captd., Ringgold, GA, POW Rock Island, IL to 22 Apr '64; rejected for USN; NFR.

Westmoreland, Robert P. 8 May 1862, GA; pvt.; Apr/Jul '62 and Jul/Aug '62 AWOL; deserted 15 Jun '62; 23 Jul '62 enl. Co. H, 65th GA INF; 1860 Buncombe Co., NC, #272, 22 y/o, 1 child; b. 28 May 1838, d. 22 Dec. 1911, bur. Friendship Cem., Cale, Nevada Co., AR. All Westmorelands are brothers.

Westmoreland, Seaborne L. (Christopher L) 16 May 1862, GA; pvt.; Apr/Jul '62 pres.; Jul/Aug '62 AWOL 26 Jul; Sep/Oct '62 deserted 4 Aug '62; 16 Aug '62 enl. Co. K, 65th GA; 1860 Habersham, GA, #623, 19 y/o; b. 6 Nov 1840, d. 28 Nov 1935, bur. Rose Hill Cem., Royston, Franklin Co., GA. All Westermorelands are brothers.

Westmoreland, Thomas .H. 16 May 1862; GA; pvt.; Apr/Jul '62 AWOL 15 Jun; Jul/Aug '62 AWOL 15 Jun; Sep/Oct '62 deserted 21 Jun '62; 21 Jun '62 enl. Co. K, 65th GA INF; 1860 Habersham, GA,

#629, 25 y/o, five children; b. 25 Jul 1829, d. 1899, bur. Hall Co., GA. All Westmorelands are brothers.

White, William H. 24 Feb 1862; GA; pvt.; 4 Jun '62 dc'd for chronic inflammation of kidneys and bladder; b. 1826, Pickens, SC; NFR.

Whitworth, Joseph 12 May 1862; GA; pvt.; Apr/Jul '62 pres.; Jul/Aug '62 AWOL 26 Jul; Sep/Oct '62 des. 4 Aug '62; NFR.

Williams, Alex 14 Mar 1863; Darlington, SC; pvt.; Mar/Apr '63 sick in camp; May/Jun '63 hosp. since Apr; Nov/Dec '63 hosp. since Apr; NFR.

Williams, Henry 11 Jun 1862; Charleston, SC; pvt.; Apr/Jul '62 pres.; Jul/Aug '62 AWOL 26 Jul; Sep/Oct '62 des. 21 Jul; transf. to navy by 30 Sept '62; NFR.

* **Wilson, James M.** 7 Jun 1863; Pickens, SC; pvt.; Nov/Dec '63 pres.; d. hosp. in Dalton, GA, claim by widow, Elizabeth 14 Nov. '64; tentative 1860 Pickens, SC, Walhalla, #571, presumed wife "Mary" and one child.

Wilson, Thomas 12 Feb 1862; New Orleans, LA; cpl.; Apr/Jul '62 pres., apptd. cpl. 1 Jun; Jul/Aug '62 pres.; Sep/Oct '62 pres.; Nov/Dec '62 pres.; Jan/Feb '63 pres., pvt. 1 Feb; Mar/Apr '63 pres.; May/Jun '63 pres.; Nov/Dec '63 pres.; May/Jun '63 transf. to Navy; NFR.

Winningham, Isaac 29 Jan 1862; Charleston, SC; pvt.; 4 Jun '62 disability discharge for hemoptysis and necrosis of lower jaw; b. 1838 St. Paul's Parish, Charleston; NFR.

Wood, Walker G. 10 Feb 1862; Huntsville, AL; pvt.; Apr/Jul, Jul/Aug, Sep/Oct, Nov/Dec '62 pres.; Jan/Feb, Mar/Apr '63 pres.; cpl. 15 Jun. '63; Nov/Dec '63 pres.; clothing 19 Apr and 30 Sep '64; 1860 Madison Co., Huntsville, AL, #199, 30 y/o and wife; d. 23 Apr 1891, bur. Maple Hill Cem., Huntsville, AL; VA headstone provided by author.

* **Wooten, Jasper N.** 18 Apr 1862; Pickens, SC; pvt.; Jul/Aug '62 pres.; Sep/Oct '62 pres.; Nov/Dec '62 abs. sick furlough 29 Nov; Jan/Feb, Mar/Apr and May/Jun '63 pres.; Nov/Dec '63 pres; d. Dalton, GA before father's claim 29 Feb '64.

* **Wooten, John W.** 15 Feb 1862; Pickens, SC; pvt.; Apr/Jul '62 pres.; Jul/Aug '62 abs., sick Summerville, SC hosp.; Sep/Oct and Nov/Dec '62 pres.; d. 20 Feb '63 Camp Helen, Mt. Pleasant, SC, claim filed 28 Apr '63 by widow, Emeline A.; tentative 1860 Anderson, SC, Regiment 2, #227, "James" and Emaline, one child, no ages.

Wynn, D.H. 21 Feb 1863; Columbia, SC; pvt.; Nov/Dec '63 pres.; clothing 7 May, 27 Jun and Aug '64; NFR.

Yearwood, Singleton Sisk 17 May 1862, GA; pvt.; Apr/Jul '62 pres.; Jul/Aug '62 AWOL 20 Jun; Sep/Oct '62 abs., des. 20 Jun; 1860 Habersham, GA, #557, 14 y/o; d. 27 Dec 1929, bur. Old Toccoa Ch., Stephens Co., GA.

Young, Abner M. 7 Apr 1863, Pickens, SC; pvt.; Mar/Apr '63 not yet reported; May/Jun to Nov/Dec '63 pres.; Mar and Jun '64 clothing; 28 Jul severe groin wnd., Kennesaw Mtn., GA; Aug '64 clothing; 30 Oct '64 captd. Decatur, AL, POW Camp Douglas, IL to 15 Apr '65 enl. in 5th US Volunteers, mustered out 13 Nov '66, cook; 1860 Pickens, SC, Regt. 5, #1142, 18 y/o; b. Mar 1846, d. Mar 1905, bur. Flat Rock Bapt. Ch., Liberty, Pickens, SC. Brother of William Jasper.

* **Young, Archibald M.** 25 Mar 1862, Pickens, SC; pvt.; Apr/Jul '62 to May/Jun '63 pres., extra duty shoeing horses; 25 Nov '63 wnd. and MIA at Missionary Ridge, TN; 26 Nov captd., Ringgold, GA; 1 Feb '64 admitted to US hosp., Chattanooga, TN, GSW left thigh; d. 2 Mar '64, fracture thigh and diarrhea, US Gen Hosp. 1, Nashville, TN; b. 28 Sep 1826, Pickens, 32 y/o, married; bur. #6535 Nashville, TN City Cem.; 1860 Pickens, SC, Regt. 5, #880, 32 y/o blacksmith, 4 children.

Young, William Jasper 14 Apr 1861 enl. 4th SC INF, Pickens, SC, 1 year, dc'd. for paralysis 4 Oct '61; 6 Oct '63, Columbia, SC, enl. FAC; pvt.; Apr, Jul and Aug '64 clothing; 12 Apr 1865 captd., Salisbury, NC, POW Camp Chase, OH to 13 Jun '65; 1860 Pickens, SC, Regt. 5, #165, 27 y/o, 1 child; b. 17 Nov 1833, d. 30 Dec 1908, bur. Corinth #2 Bapt. Ch., Anderson, SC. Brother of Abner M.

Notes

Chapter 1. Captain Thomas B. Ferguson

1. John Peyre Thomas, *The History of the South Carolina Military Academy* (Charleston: Walker, Evans & Cogwell, 1893), 107.

2. H.M. Clarkson, "Story of the Star of the West," *Confederate Veterans Magazine* (May, 1913), 21:236; A.G.D. Wiles, "The Boys Behind the Gun - The Citadel Cadets Who Fired the First Shot on the *Star of the West,*" *The Congressional Record Appendix* (4 January 1961), A22. Clarkson and Wiles agree that Ferguson, on gun #2, hit the ship with his shot.

3. *Reports and Resolutions of the General Assembly of the State of South Carolina vol. 1, Annual Report of the State Historian of Confederate Records for the Year 1899* (Columbia: Bryant Printing Co.), 11.

4. Arthur Brailsford Wescoat, "Journal of Arthur Brailsford Wescoat 1863, 1864," *South Carolina Historical Magazine* 55 (1954), 81. Entries for 15 to 18 July 1863 mention the deaths of brothers Joseph and Thomas Ferguson, but mistakenly state that Thomas received a mortal wound at Morris Island, SC in the attack on Battery Wagner, rather than his non-fatal wound at Jackson, MS.

5. *Charleston (SC) Mercury,* September 24, 1864, 1; November 25, 1864, 2. (Hereafter cited as *CM*).

6. *National Cyclopedia of American Biography* (New York: James T. White & Co, 1906), 31:343.

7. Office of the Historian, US Department of State, https://history.state.gov/departmenthistory/people/ferguson-thomas-barker.

8. Elizabeth S. Boyd, *Boyd's Directory of the District of Columbia 1912* (Washington, DC: R.L. Polk & Co.), 590; 1900 US census, population schedule, Washington, DC, district 33, family # 230.

9. Thomas B. Ferguson, Inverting coffee-making apparatus. US Patent 782956, filed January 20, 1904, and issued February 21, 1905; *Official Gazette of the United States Patent Office*, January 22, 1878, 155-56, No. 199,527. The latter shows drawings of Ferguson's Fish-Hatching Apparatus. https://archive.org/details/officialgazette13unit/page/n86/mode/1up

10. "Thomas Barker Ferguson," https://www.findagrave.com/memorial/170399605/thomas-barker-ferguson.

Chapter 2. The Officers

1. 1900 US census, St. Bernard Parish, LA, ward 1, district 45, family #165. However, the censuses of 1850, 1860 and 1880 suggest a birth year of 1841, and the 1870 census suggests 1842.

2. René T. Beauregard, "Reminiscence," undated, G.T. Beauregard Miscellany, 5, 83.

3. Germaine M. Reed, *David French Boyd Founder of Louisiana State University* (Baton Rouge: LSU Press, 1977), 41.

4. *Combined Service Records of Confederate Soldiers Who Served In Organizations from the State of South Carolina.* (hereafter cited as *CSR).* Beauregard has entries in both the "South Carolina" and the "Officers" sections, since he served on his father's staff at war's end. Letters from his senator and congressman in LA, the latter family with whom he lived as a child, sought his premature commission unsuccessfully.

5. Judge René Toutant Beauregard, https://www.findagrave.com/memorial/70585491.

6. Robin Jaffee Frank, *Love and Loss American Portrait and Mourning Miniatures* (New Haven: Yale University Press, 2000), 151.

7. 1860 US census, Lower All Saints Parish, Georgetown, SC, div. 3, family # 809.

8. 1910 US census, Sumter Co., Rafting Creek, SC, Distr. 114, family #183. Emma claims three children, two living.

9. Death Certificate, Charleston, SC Health Department for Dr. Jno Ashe Alston.

10. 1850 US census, Pickens, Eastern Division, SC, family #869.

11. 1860 US Agricultural Census, Pickens, SC.

12. 1860 US Slave Schedule, Pickens, Regiment 5, SC, 22-23; Klosky. *The Pendleton Legacy* (Columbia: Sandlapper Press, 1971). Cited by Lynn R. Beeson. *In the Face of Change,* https://www.clemson.edu/public/experimental-forest/documents/prospectus.pdf.

13. *Anderson Intelligencer,* October 12, 1865.

14. Ernest M. Lander, Jr., *The Calhoun Family and Thomas Green Clemson: the decline of a southern patriarchy* (Columbia: USC Press, 1983), 155.
15. Clement Anselm Evans, ed., *Confederate Military History: A Library of Confederate States History, 10* (Atlanta: Confederate Publishing Co., 1899), 447.
16. Joseph Henry Anatole Hincks, https://www.findagrave.com/memorial/70632944.
17. 1850 US census, Parish of St. Phillips and St. Michael's, Charleston, SC, family #1146.
18. California, Voter Registers, Merced County, 1867-1898, https://www.ancestry.com/.
19. Corp. Ion Simons, https://www.findagrave.com/memorial/27597559.
20. 1850 US census, Orangeburg, SC, family #554.
21. *Biographical and Historical Memoirs of Pulaski, Jefferson, Lanoke, Faulkner, Grant, Saline, Perry, Garland and Hot Spring Counties, Arkansas* (Goodspeed Publishing, 1889), 504.
22. Dr. Louis Raoul Stark, https://www.findagrave.com/memorial/7840521.
23. William Izard Bull. Death Certificate, Charleston, SC; 1860 US Slave Schedule, Charleston, Charleston City, SC, Ward 2, https://www.ancestry.com/.
24. Arthur Wayne Hafner, *Directory of Deceased Physicians, 1804-1929* (Chicago: AMA, 1993), https://www.ancestry.com/search/collections/7833/.
25. Deering J. Roberts, MD, "Organization of the Confederate Medical Department," *The Photographic History of the Civil War, 4*, http://www.medicalantiques.com/civilwar/Civil_War_Articles/Confederate_Army_Surgeons.
26. Bull, Death Certificate.

Chapter 3. Recruiting the Company

1. United States War Department. *The War of the Rebellion: A Compilation of the Official Records of the Union and Confederate Armies* series 4, volume 1, sec. 1, 128. (Hereafter cited as *OR*), https://babel.hathitrust.org/cgi/pt?id=coo.31924080778909&view=1up&seq=1.
2. James M. Matthews, ed. *Statutes at Large of the Confederate States of America, First Session of the First Congress; 1862* (Chapel Hill: UNC, 1999), chap. 31, 32, April 16, 1862, https://docsouth.unc.edu/imls/statutes/statutes.html.
3. James A. James to his parents, August 20, 1863. A cousin, Griffin James, stayed in Co. E, 1st SC, Butler's Regt, and was killed 8 Sept. 1863 in the bombardment of Fort Moultrie, SC, http://historicalshop.com/sitecontents/confederate/document.htm.
4. *OR*, ser. 4, vol. 2, 816.
5. *OR*, ser. 4, vol. 1, sec. 1, 129-30.
6. Confederate States of America. War Department. *Regulations for the army of the Confederate States, 1862* (Richmond: West & Johnson, 1862), 104, https://archive.org/details/regulationsforarconf/page/6/mode/2up.

7. *OR*, ser. 4, vol. 1, sec. 1, 129-30.
8. William Sizemore, alias William Seymore, Declaration for Invalid Pension and associated forms, NARA, Certificate # 797107.
9. Samuel Lorenzo Malone, *The Sizemore Family Tree: One Man's Search for his roots* (San Francisco: Diego's Printing, 1980), 49, 65.
10. Malone, "Sizemore Family Tree (Pickens, SC) 1830-1900." Page not numbered, follows Contents page.
11. *CSR*, William Sizemore, 5th SC INF, 12.
12. Sizemore, pension application.
13. Malone, *Sizemore Family Tree,* 49.
14. Malone, *Sizemore Family Tree,* appendix, Bureau of Indian Affairs to Samuel L. Malone, Aug. 9, 1978. No Cherokee man in the database named William Sizemore matched Malone's ancestor.
15. Matthews, *Statutes*, chap. 45.
16. 1860 US census, Charleston, St. Johns Berkeley Parish, SC, families #139, 143.
17. David Dangerfield, e mail exchange and telephone conversation with author April 19, 2020.
18. Laura A. Huff, e mail communication to author; Dangerfield, April 19, 2020.
19. Confederate States of America, War Department, *Articles of war, for the government of the Army of the Confederate States* (Montgomery: Barrett, Wimbish & Co., 1861), art. 22, 6, https://archive.org/details/articlesofwarfor00conf/page/5/mode/2up.

Chapter 4. Camp Life: Crime and Punishment

1. Louise Matheson Bell, ed., *Soldiers in Grey, Rebels from Pickens District 1861 – 1865* (Clemson, SC: Joyce's Print Shop, 1984).
2. Confederate States of America, *Articles of war*, art. 21, 6.
3. Confederate States of America, *Articles of war*, art. 20.
4. Confederate States of America, *Articles of war*, art. 23, 7.

Chapter 5. The Weapon

1. Confederate States of America, *Regulations for the army, 7.*
2. T.T.S. Laidley, *The Ordnance Manual for the Use of the Officers of the Unites States Army, 3rd ed.*, 1861 (Decatur, MI: INVICTUS, 2003), 280.
3. Laidley, *Ordnance Manual,* 280.
4. Laidley, *Ordnance Manual,* 335.
5. James Morgan, "Mounted But Not Mounted" The Confusing Terminology of Artillery, http://vcwsg.com/PDF%20Files/Mounted%20But%20Not%20Mounted.pdf.

6. *OR*, ser. 1, vol. 32, pt. 3, 684.
7. *OR*, ser. 1, vol. 32, pt. 3, 694.
8. George Patten, *Artillery Drill: containing Instructions in the School of the Piece and Battery Manoeuvres* (New York: J.W. Fortune, 1864), 17. Videos of the living historians of Ferguson's SC Artillery Co. utilizing this drill can be found at https://www.youtube.com/watch?v=GZapd63A1zs).
9. Confederate States of America, *Regulations for the army*, 104.

Chapter 6. Supplying the Company

1. Confederate States of America. War Department, and Confederate States of America Collection. *Uniform and dress of the army of the Confederate States* (Richmond: Wynne, 1861), 3, 5, https://www.loc.gov/item/05025942/. Jackets issued to the men were apparently not "artillery" in style, prompting Ferguson, ever mindful of regulations, to request the proper red trim. In 1862 a blue-trimmed red cap replaced a red-trimmed gray one.
2. Laidley, *Ordnance Manual*, 134. A prolong is a stout rope coiled on the gun carriage for dragging/towing in emergencies.

Chapter 7. Losses: Casualties, Captures, and Sickness

1. 1860 US census, Population Schedule, Heard Co., GA, family #483; 1860 US census, Slave Schedule, Heard Co., GA, owner John Benefield.
2. Carl Alexander Retzloff, "'The First Fruits of a Re-united People' The Loyalty, Motivation and Allegiance of the Men of the Second United States Volunteer Infantry Regiment" (honors thesis, Washington and Lee University, 2015), 14.
3. Retzloff, "First Fruits," 3.

Chapter 8. The Campaign at Jackson, MS

1. *OR*, ser. 1, vol. 14, 923.
2. *OR*, ser. 1, vol. 14, 924.
3. *OR*, ser. 1, vol. 14, 925.
4. *OR*, ser. 1, vol. 14, 925, 926; Ellison Capers, *Confederate Military History vol. 5 South Carolina* (Secausus, NJ: The Blue & Grey Press), 203.
5. *OR*, ser. 1, vol. 24, pt. 3, 833.
6. *OR*, ser. 1, vol. 14, 948.
7. *OR*, ser. 1, vol. 31, pt. 2, 99.
8. Robert S. Seigler, *South Carolina's Military Organizations During the War Between the States: The Lowcountry & Pee Dee* (Charleston: The History Press, 2008), 275.
9. *OR*, ser. 1, vol. 24, pt. 3, 883.
10. *OR*, ser. 1, vol. 24, pt. 3, 919.

11. *OR,* ser. 1, vol. 24, pt. 3, 920.
12. Capers, *Military History*, 211.
13. Bell, *Rebels in Grey,* 81. The CSR shows this soldier died 10 Feb. 1863, before this letter was written; the conflict in dates is unresolved.
14. Jim Woodrick, *The Civil War Siege of Jackson, Mississippi* (Charleston, SC: History Press, 2016), 52.
15. Woodrick, *Siege of Jackson,* 58.
16. Woodrick, *Siege of Jackson,* 63-4.
17. Flavel C. Barber, *Holding the Line: The Third Tennessee Infantry 1861-1864,* ed. Robert H. Ferrell (Kent, OH: Kent State University, 1994), 137.
18. John McMillan Armstrong to sister, July 14, 1863, Grace Armstrong Coile Collection, TN State Library and Archives, Nashville.
19. *CM,* July 16, 1863, 2, col. 1.
20. Woodrick, *Siege of Jackson,* 95-6.
21. *OR*, ser. 1, vol. 24, pt. 3, 1008.
22. Woodrick, Siege of *Jackson,* 100-102.

Chapter 9. Tennessee in '63

1. *OR*, ser. 1, vol. 30, pt. 4, 516.
2. *OR*, ser. 1, vol. 30, pt. 4, 538.
3. *OR*, ser. 1, vol. 30, pt. 4, 541.
4. *OR*, ser. 1, vol. 30, pt. 4, 594-95.
5. *OR*, ser. 1, vol. 30, pt. 2, 11.
6. *OR*, ser. 1, vol. 30, pt. 2, 244; Capers, *Military History,* 277, 280.
7. James A. James to John James, October 20, 1863, Spared & Shared 9, https://sparedshared9.wordpress.com/2015/11/28/1864-james-a-james-to-john-james/.
8. *OR*, ser. 1, vol. 32, pt. 2, 686.
9. *OR*, ser. 1, vol. 31, pt. 2, 722.
10. Col. James Cooper Nisbet, *Four Years on the Firing Line (Chattanooga: the Imperial Press, 1914),* 248-49.
11. *OR*, ser. 1, vol. 31, pt. 2, 747.
12. USGS topographic maps, https://viewer.nationalmap.gov/basic.
13. Nisbet, *Four Years,* 255.
14. *OR* ser. 1, vol. 31, pt. 2, 657, 679.
15. John Berrien Lindsley, *The Military Annals of Tennessee. Confederate. First series* (Nashville: J. M. Lindsley & Co., 1886), 217, https://archive.org/details/cu31924032778700/page/216/mode/2up.
16. *Covington (TN) Leader*, Nov. 26, 1863.
17. *OR*, ser. 1, vol. 31, pt. 2, 480.
18. *National Tribune, Washington, D.C.*, 11 Feb 1892, 4, col. 2.
19. *Yorkville (SC) Enquirer*, 9 Dec 1863, 1, col. 3.

20. G.T. Beauregard to Capt. A.N.T. Beauregard, Dec. 22, 1863. War Dept. Collection of Confederate Records, R.G. 109, Dept. of SC, GA and FL, Letter Book No. 2, chap. 2, no. 183, 164.

21. Bull and Young to Margaret Young, collection of Robert Archibald Young, III, Fort Smith, AR.

Chapter 10. The Atlanta Campaign

1. *OR*, ser. 1, vol. 32, pt. 3, 685.
2. *OR*, ser. 1, vol. 32, pt. 3, 687.
3. *CSR*, Lt. Col. Hypolite Oladowski.
4. *OR*, ser. 1, vol. 32, pt. 3, 693-94.
5. *OR*, ser. 1, vol. 32, pt. 2, 645.
6. *OR*, ser. 1, vol. 32, pt. 2, 817.
7. *OR*, ser. 1, vol. 32, pt. 2, 698.
8. *CSR*, Oladowski.
9. *OR*, ser. 1, vol. 32, pt. 3, 688.
10. Walter Brian Cisco, *States Rights Gist A South Carolina General of the Civil War* (Gretna, LA: Pelican Publishing, 2008), 116.
11. James A. James to John James and mother, January 18, 1864. Spared & Shared 9, https://sparedshared9.wordpress.com/2015/11/28/1864-james-a-james-to-john-james/.
12. James A. James to parents, April 27, 1864. Item #14250, http://historicalshop.com/sitecontents/confederate/document.htm.
13. *OR*, ser. 1, vol. 38, pt. 3, 615.
14. *CM,* Nov. 1, 1864, 1.
15. *CM,* Nov. 1, 1864, 1.
16. *OR*, ser. 1, vol. 38, 33, 378.
17. *OR*, ser. 1, vol. 38, 715.
18. "Battle of Rome Cross Roads," https://en.wikipedia.org/wiki/Battle_of_Rome_Cross_Roads.
19. Sam R. Watkins, *Co. Aytch* (New York: Touchstone, 1997), 154.
20. James A. James to John James, June 20, 1864, Spared & Shared, https://sparedshared9.wordpress.com/2015/11/28/1864-james-a-james-to-john-james/.
21. *CM,* November 1, 1864, 1, col. 5.
22. Richard M. McMurry, *Atlanta 1864 Last Chance for the Confederacy* (University of Nebraska Press, 2000), 109.
23. Watkins, *Co. Aytch,* 156. The infantry fight at the "Dead Angle" is described in dramatic detail.
24. Watkins, *Co. Aytch,* 178.

25. *OR*, pt. 1, vol. 38, 631.
26. Cisco, *States Rights Gist,* 129.
27. *OR*, pt. 3, vol. 38, 633.
28. *CM,* November 1, 1864

Chapter 11. Back to Tennessee

1. *OR* ser. 1, vol. 45, pt. 1, 738.
2. *OR* ser. 1, vol. 39, pt. 1, 826.
3. Capers, *Military History,* 340.
4. René Beauregard to J.A. Chalaron, October 4, 1875, "A brief history of Ferguson's battery with reminiscences of the Tennessee Campaign, Louisiana Historical Association Collection, Tulane University.
5. *OR*, ser. 1, vol. 45, pt. 1, 683–84.
6. *OR*, ser. 1, vol. 39, pt. 3, 888.
7. *OR*, ser. 1, vol. 39, pt. 1, 827.
8. *CSR*, Confederate Staff Officers, Major Robert Cobb, Cobb's Bttn. Artillery; Beauregard to Chalaron, Oct. 4, 1875.
9. *OR* ser. 1, vol. 45, pt. 1, 652.
10. *OR*, ser. 1, vol. 45, pt. 1, 743.
11. *OR*, ser. 1, vol. 45, pt. 1, 653.
12. David R. Logsdon, comp. and ed., *Eyewitnesses at the Battle of Franklin* (Kettle Mills Press, 2000), 83.
13. *OR*, ser. 1, vol. 45, pt. 1, 731, 747.
14. Nathaniel Cheairs Hughes, Jr., *The Pride of the Confederate Artillery, The Washington Artillery in the Army of Tennessee* (LSU Press, 1997), 246; Beauregard to Chalaron, Oct. 4, 1875.
15. Hughes, *Pride of the Artillery*, 536.
16. Hughes, *Pride of the Artillery*, 539.
17. *History of Lafayette County, MO* (St. Louis, 1881), 371.
18. *OR*, ser. 1, vol. 45, pt. 1, 748.
19. Hughes, *Pride of the Artillery,* 250, 251.
20. *OR* ser. 1, vol. 45, pt. 1, 749.
21. *OR* ser. 1, vol. 45, pt. 1, 747-51.
22. *OR* ser. 1, vol. 45, pt. 1, 478, 479.
23. Hughes, *Pride of the Artillery*, 251.
24. *OR* ser. 1, vol. 45, pt. 1, 732.
25. Beauregard to Chalaron, Oct. 4, 1875.

Chapter 12. The Carolinas Campaign

1. "Sketches of the Third Maryland Artillery," *Southern Historical Society Papers v. 11, no. 10,* October 1883.

2. Lindsley, *Military Annals,* 835-37.
3. *OR*, ser. 1, vol. 47, pt. 3, 756.
4. *OR*, ser. 1, vol. 49, pt. 1, 333.
5. Ann Brownlee, “Stoneman’s Raid: Salisbury and the Yadkin River Bridge,” www.trading-ford.org/stoneman.html.
6. Cornelia Phillips Spencer, *The Last Ninety Days of the War in North Carolina* (New York: Watchman Publishing, 1866), 200.
7. Lindsley, *Military Annals*, 838-40.
8. *OR*, ser. 1, vol. 49, pt. I, 334.
9. Spencer, *Last Ninety Days*, 218.
10. Lindsley, *Military Annals*, 842.

Bibliography

Ancestry.com.

Anderson (SC) Intelligencer, https://www.loc.gov/item/sn84026965/?st=calendar.

Baker, Gary R. *Cadets in Gray*. Columbia, South Carolina: Palmetto Bookworks, 1989.

Barber, Flavel C. *Holding the Line: The Third Tennessee Infantry 1861-1864.* Edited by Robert H. Ferrell. Kent, OH: Kent State University, 1994.

Beauregard, René T. "A brief history of Ferguson's battery with reminiscences of the Tennessee Campaign." letter to J.A. Chalaron. October 4, 1875. Tulane University Library.

Beauregard, René T. "Reminiscence." n.d. G. T. Beauregard Miscellany. Louisiana State University Library.

Bell, Louise Matheson, ed. *Rebels in Grey, Soldiers from Pickens District 1861-1865.* Clemson, South Carolina: Joyce's Print Shop, 1984.

Biographical and Historical Memoirs of Pulaski, Jefferson, Lanoke, Faulkner, Grant, Saline, Perry, Garland and Hot Spring Counties, Arkansas. Goodspeed Publishing, 1889. https://archive.org/details/biographicalhistpjlf00good/page/n5/mode/2up.

Bibliography

Boyd, Elizabeth S. *Boyd's Directory of the District of Columbia 1912.* Washington, DC: R.L. Polk & Co., 590 https://catalog.hathitrust.org/Record/000527013.

Brownlee, Ann. "Stoneman's Raid: Salisbury and the Yadkin River Bridge."www.trading-ford.org/stoneman.html

Busha, Charles H., comp. *They Fell to the Southern Cause Confederates of the Eastern District of Pickens.*

Capers, Brig. Gen. Ellison. *Confederate Military History vol. 5 South Carolina.* Edited by Clement A. Evans. The Blue & Grey Press.

Chambers, Herbert O. III. *And were the glory of their times, artillery.* Wilmington, NC: Broadfoot Publishing Co. 2014.

Charleston (SC) Mercury.

Chastain, James Garvin D.D. *The Chastain Family Tree.* n.d.

Chastain, Victor T. *Chastains of Pickens County, SC 1790-1986 Ten Shilling Bell.* n.d.

Cheek, Linda, comp. *Pickens District, SC 1860 Census 2nd and 5th Regiments.*

Cisco, Walter Brian. *States Rights Gist A South Carolina General of the Civil War.* Gretna, LA: Pelican Publishing, 2008.

Combined Service Records of Confederate Soldiers Who Served In Organizations from the State of South Carolina, National Archives and Records Administration. https://www.fold3.com.

Confederate Pension Rolls Oconee Co. 1888, Comptroller General, RG 126, SC Archives.

Confederate Pension Rolls Pickens Co. 1891, Comptroller General, RG 126, SC Archives.

Confederate States of America. War Department. *Articles of war, for the government of the Army of the Confederate States.* Montgomery: Barrett, Wimbish & Co., 1861. Art. 22, 6. https://archive.org/details/articlesofwarfor00conf/page/5/mode/2up.

Bibliography

Confederate States of America. War Department. *Regulations for the army of the Confederate States, 1862.* Richmond: West & Johnson, 1862, 104. https://archive.org/details/regulationsforarconf/page/6/mode/2up.

Confederate Veteran. 40 vols. Nashville, TN.

Daniel, Larry J. *Cannoneers in Gray The Field Artillery of the Army of Tennessee Revised Edition.* Tuscaloosa: The University of Alabama Press, 2005.

Dedmondt, Glenn. *The Flags of Civil War South Carolina.* Pelican, 2000.

Evans, Clement Anselm, ed. *Confederate Military History: A Library of Confederate States History. 10.* Atlanta: Confederate Publishing Co., 1899.

Ferguson, Thomas B. Inverting coffee-making apparatus. US Patent 782956, filed January 20, 1904, and issued February 21, 1905.

Find A Grave, https://www.findagrave.com/.

Frank, Robin Jaffee. *Love and Loss American Portrait and Mourning Miniatures.* New Haven: Yale University Press, 2000.

Hafner, Arthur Wayne. *Directory of Deceased Physicians, 1804-1929.* Chicago: AMA, 1993. https://www.ancestry.com/search/collections/7833/.

History of Lafayette County, MO. St. Louis: Missouri Historical Company, 1881.

Hitch, Michael G. and Elinor Mowbray. *Oh Lordy!* Outskirts Press, Inc. 2015.

Holcomb, Brent H. *South Carolina's Confederate Pensioners in 1901.* SCMAR 2001.

Hughes, Nathaniel Cheairs Jr., *The Pride of the Confederate Artillery, The Washington Artillery in the Army of Tennessee.* LSU Press, 1997.

Keowee (Pickens Court House, SC) Courier. https://chroniclingamerica.loc.gov/lccn/sn84026912/issues/.

Klosky, Beth Ann. *The Pendleton Legacy.* Columbia: Sandlapper Press, 1971. Cited by Lynn R. Beeson. *In the Face of Change.* https://www.clemson.edu/public/experimental-forest/documents/prospectus.pdf.

Knudsen, Lewis F. Jr, comp. *Wayside (1st Louisiana) Confederate Hospital Charleston, SC 1862-1865 Patient Register* 1998.
Laidley, T.T.S. *The Ordnance Manual for the Use of the Officers of the United States Army 3rd ed.* 1861. Decatur, Michigan: INVICTUS, 2003.
Lander, Ernest M. Jr. *The Calhoun Family and Thomas Green Clemson: the decline of a southern patriarchy*. Columbia: USC Press, 1983.
Leishman, Robert. *Pierre Chastain and His Descendants Vol. I.* Pierre Chastain Family Association, 1995.
Lindsley, John Berrien. *The Military Annals of Tennessee. Confederate. First series.* Nashville: J. M. Lindsley & Co., 1886.
Logsdon, David R. comp. and ed. *Eyewitnesses at the Battle of Franklin.* Kettle Mills Press, 2000.
Malone, Samuel Lorenzo. *The Sizemore Family Tree: One Man's Search for his roots.* San Francisco: Diego's Printing, 1980.
Matthews, James M., ed. *Statutes at Large of the Confederate States of America, First Session of the First Congress*; *1862.* Chapel Hill: UNC, 1999. https://docsouth.unc.edu/imls/statutes/statutes.html.
McCuen, Anne K. *Including a Pile of Rocks.* Greenville, SC: Southern Historical Press, Inc. 2005.
McMurry, Richard M. *Atlanta 1864 Last Chance for the Confederacy.* University of Nebraska Press, 2000.
Melton, Jack W. Jr. *Civil War Artillery Projectiles The Half Shell book.* Charleston, SC: Historical Publications LLC, 2018.
Message No. 1 of his Excellency Gov. J.L. Orr with Accompanying Documents July 1868, Columbia, SC.
Militia Enrollments List of Men in Eastatoe, Pickens (SC) 30-45 years old, 1869, online SC Dept. of Archives and History.
Morgan, James. "Mounted But Not Mounted" The Confusing Terminology of Artillery. http://vcwsg.com/PDF%20Files/Mounted%20But%20Not%20Mounted.pdf.
National Cyclopedia of American Biography, The. vol. 31 New York: James T. White & Co, 1906.

Nisbet, Col. James Cooper. *Four Years on the Firing Line.* Chattanooga: The Imperial Press, 1914.

Office of the Historian, US Department of State. https://history.state.gov/departmenthistory/people/ferguson-thomas-barker.

Onofrio, Jan. *South Carolina Biographical Dictionary.* 2000.

Patten, George. *Artillery Drill: School of the Piece and Battery Manoeuvres.* New York: J.W. Fortune, 1864.

Pickens County Board of Honor Confederate Pension Applications 1914-1933 microfilm L 39033.

Pickens County (SC) Board of Honor 1922-1950 microfilm L 39034.

Pickens County Confederate 150 Project vol. 1 2012 and vol. 2 2013.

Recollections and Reminiscences 1861-1865 through World War I. South Carolina Division United Daughters of the Confederacy, 1990.

Reed, Germaine M. *David French Boyd Founder of Louisiana State University.* Baton Rouge: LSU Press, 1977.

Reports and Resolutions of the General Assembly of the State of South Carolina vol. 1. Annual Report of the State Historian of Confederate Records for the Year 1899. Columbia: Bryan Printing Co., 1900.

Retzloff, Carl Alexander. "'The First Fruits of a Re-united People' The Loyalty, Motivation and Allegiance of the Men of the Second United States Volunteer Infantry Regiment." (Honors Thesis, Washington and Lee University, 2015).

Roberts, Deering J., MD. "Organization of the Confederate Medical Department," *The Photographic History of the Civil War, v.4.* Last modified July 3, 2019, http://www.medicalantiques.com/civilwar/Civil_War_Articles/Confederate_Army_Surgeons.

Samuel Wragg Ferguson Papers, Special Collections, Hill Memorial, LSU Libraries, Baton Rouge, LA.

SC Archives RG 126 Claims of Disability, 1886, Pickens County.

Sheriff, G. Anne, Confederate Soldiers Buried at Mt. Zion Cemetery, Central, SC.

Spencer, Cornelia Phillips. *The Last Ninety Days of the War in North Carolina.* New York: Watchman Publishing, 1866.

Thomas, John Peyre. *The History of the South Carolina Military Academy.* Charleston: Walker, Evans & Cogswell, 1893. https://catalog.hathitrust.org/Record/008591723.

SC Historical and Genealogical Magazine vol. 7, 1906 ; vol. 102, 2001, No. 1 and No. 4.

Southern Historical Society Papers.

SC Archives S190025 Dept. of Mental Health SC State Hospital Admission books: Patients Nos. 1-2190.

SC Archives S190057 Orders for Commitment and Detention of Patients 1828 to 1877 SC Lunatic Asylum.

SC Archives "Personal History and Case Books Male and Female Departments Sept. 5 1859 to Aug. 1869.

Uniform and Dress of the Army of the Confederate States. Richmond, VA, 1861. https://www.loc.gov/item/05025942/.

United States census. http://www.ancestry.com.

United States War Department. *The War of the Rebellion: A Compilation of the Official Records of the Union and Confederate Armies.* 128 vols. Washington, DC: Government Printing Office, 1891-1895.

Watkins, Sam R. *Co. Aytch.* New York: Touchstone, 1997.

Wescoat, Arthur Brailsford. "Journal of Arthur Brailsford Wescoat 1863, 1864," South *Carolina Historical Magazine* 55 (1954).

Wiles, A.G.D. "The Boys Behind the Gun - The Citadel Cadets Who Fired the First Shot on the *Star of the West.*" *The Congressional Record Appendix* (4 January 1961).

Made in the USA
Las Vegas, NV
03 November 2026

57423721R00102